Study Guide

Anthony J. Gaudin

Ivy Tech State University

to accompany

The Sciences
An Integrated Approach

FOURTH EDITION

James Trefil
George Mason University

Robert M. Hazen
George Mason University

WILEY

JOHN WILEY & SONS, INC.

To order books or for customer service call 1-800-CALL-WILEY (225-5945).

ISBN 0-471-44918-0

Printed in the United States of America

10 9 8 7 6 5 4 3 2 1

Printed and bound Bradford & Bigelow, Inc.

To The Student

This Study Guide is written to be used with the textbook: <u>The Sciences: An Integrated Approach</u>, Fourth Edition, by James Trefil and Robert Hazen. The study guide is not intended to be a substitute for the textbook or the lectures and discussion sections that make up important parts of your course. In order to be successful (that is, learn the most and earn the highest possible grade) in your science class, you must attend lectures, take notes, and organize and study those notes. Be assured that your instructor will test you on material that has been presented in class. When you are not present in class, you miss that material. Keep up with the reading assignments in the textbook. The chapters are easy to read and will help you to understand what is going on in lecture. If you can skim over the chapters to be covered in class prior to lecture, you will have an advantage over those students who enter the class not knowing what the topic of discussion is. The study guide will help you to integrate your lectures with the textbook. The study guide also provides you with an excellent method of preparing for the exams that will have a major role in determining your grade.

Format of the Study Guide

Organization is one of the keys to success in any field that contains a lot of material. This study guide is organized into 25 chapters that correspond to the 25 chapters in the textbook. Each study guide chapter is subdivided into several sections that facilitate your understanding of the material presented in the textbook.

- Each chapter begins with a **Chapter Review**, a short narrative indicating the general contents of the textbook chapter.

- A list of **Learning Objectives** follows. These objectives give you a concrete set of accomplishments you should be able to attain after studying the textbook chapter. Your instructor may add additional objectives, based on what is covered in the lectures.

- A list of **Key Chapter Concepts** is presented next . These concepts are organized in a sequential manner following their presentation in the textbook. A thorough review of these concepts is essential for an understanding of the main ideas presented in the chapter.

- Much of science is totally dependent on the individuals who made key discoveries at opportune times in the past. An exposition of **Key Individuals** identifies those people and links them with their important discoveries.

- Many concepts in science are linked to **Key Formulas and Equations**. This textbook does not make great demands on students to use advanced mathematical skills, however, it is necessary to use certain basic algebraic formulas to demonstrate the importance and implications of specific concepts.

- **Key Concept Maps** are presented for those ideas that can be outlined in this manner. The advantage of using this approach to analyze a concept is that it reduces the material to its bare essentials and relates them in a sequential fashion.

- **Self-Tests** of the chapter allow you to evaluate your understanding of the textbook material. Both multiple-choice and fill-in questions are included for each chapter.

- A **Crossword Quiz** is presented for the majority of the chapters. The puzzle concentrates primarily on vocabulary for the chapter, but may also include the names of prominent individuals or important concepts. I hope that these puzzles provide you with an interesting way to test yourself.

- The **Answers to Self-Test Questions** include the objective tests and a solution to the crossword puzzle.

Acknowledgements

Matt Van Hattem, from John Wiley and Sons, asked me to review the first edition of Trefil and Hazen's textbook. I became immersed in that project for several months, and enjoyed contributing to their effort. Then, Jennifer Yee, also from John Wiley and Sons , offered me the opportunity to write this study guide. She devised the format for the guide, and made sure that each element contributed to the overall goal. She encouraged my efforts and kept me on schedule. Her help was invaluable during the writing of previous editions of this work. I thank her for her dedication to the project and for keeping our efforts on target. Geraldine Osnato has been my editorial guide for this, the fourth edition of the Study Guide. David Harris, Biology Editor for John Wiley and Sons, was the administrator for the entire project.

Anthony J. Gaudin, Ph.D.
Ivy Tech State College
Columbus, Indiana
email: agaudin@ivytech.edu

CONTENTS

Chapter 1

Science: A Way of Knowing

Chapter Review

Scientists use the *scientific method*, a formal series of steps that reduces the likelihood of bias, to ask and answer questions about the entire universe. The scientific method is based on offering testable *hypotheses* to explain natural events that have been observed. Those hypotheses that withstand preliminary testing are called *theories*; and theories that hold up over the years are referred to as *laws*. Nonscientific endeavors, such as philosophy, religion, and the arts are unsuitable for study using the scientific method. Scientists have used the scientific method to develop numerous technological advances that have improved the quality of human life.

Learning Objectives

After studying this chapter, you should be able to do the following:
(Other objectives may also be assigned by your instructor.)

1. Describe the scientific method and indicate how it is used to answer questions .
2. Apply the scientific method to answering a question that you have encountered in your daily activities.
3. Distinguish between sciences and pseudosciences.
4. Differentiate between basic research and applied research.
5. Identify some of the major branches of science.
6. List some of the major sources of financial support for scientific research.

Chapter Concepts

- Science is a way of gathering information about and explaining all aspects of the universe.
- The scientific method is the standard tool used by scientists to ask and answer questions about the natural world.
- The scientific method consists of: 1) making accurate observations, 2) formulating a hypothesis (a tentative guess) to explain the observation, 3) performing experiments to test the hypothesis, 4) analyzing data gathered in the experiments, and 5) developing a general theory that predicts how similar observations can be explained in the future.
- Many areas of human activity, such as art, religion, and philosophy do not lend themselves to scientific study.
- Some areas of human interest, such as astrology, are classified as psuedosciences because, even though they claim to be scientific, they cannot be analyzed using the scientific method.

- There is a large variety of scientific specialties, and scientists performing the same kind of studies communicate with one another at scientific meetings and through publishing the results of their studies in peer-reviewed journals.
- The major branches of science include physics, chemistry, astronomy, geology, and biology.
- Basic scientific research often leads to the development of new materials and technological advances that have practical application to everyday life.
- The overwhelming monetary support for American scientific research comes from the federal government, although private industry also carries on research aimed at developing new commercial products.

Key Individuals

- Ecologist David Tilman carried out an important ecological experiment in Minnesota that studied the effects of nutrients and rainfall on the quantity and diversity of life on grassy plots of land.
- The Italian astronomer Galileo Galilei demonstrated to his contemporaries that new discoveries could be made about the universe with the aid of a telescope. He used his telescope to show them the moons of Jupiter and craters on Earth's moon.
- Dimitri Mendeleev, a Russian chemist, developed the Periodic Table of the Elements in the latter part of the nineteenth century.
- During the eighteenth century, William Harvey, an English physician, used the scientific method to discover the pattern of blood circulation in animals, including humans.

Key Concept: The Scientific Method Involves Testing Hypotheses

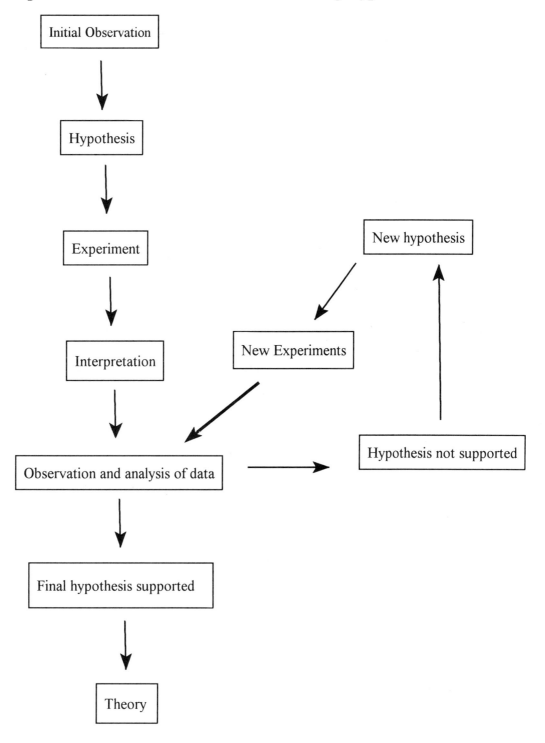

Self-Test

Multiple-Choice Questions:

1. A hypothesis is
    a. a proven solution to a problem
    b. a tentative guess offered as the solution to a problem
    c. often referred to as a law of nature
    d. the same as a theory

2. Which of the following procedures are included in the scientific method?
    a. careful observations
    b. tentative guesses
    c. analysis of data
    d. all of the above

3. A law of nature is a theory that
    a. can never be proven false
    b. can never be challenged by the scientific method
    c. has advanced above the status of a hypothesis
    d. has been tested extensively and appears to apply everywhere in the universe

4. Science is a way of producing
    a. final answers to questions
    b. ultimate truth
    c. absolutely certain solutions to problems
    d. none of the above

5. Edward Tilman's ecological experiments indicated that adding nitrogen to experimental plots of land
    a. increased the amount and variety of plants growing there
    b. decreased the amount and variety of plants growing there
    c. increased the amount but decreased the variety of plants growing there
    d. decreased the amount but increased the variety of plants growing there

6. The English physician William Harvey is renowned for his work involving
    a. the periodic table of the elements
    b. circulation of the blood
    c. the effects of nitrogen on plant growth
    d. none of the above

7. Using his own telescope, the Italian astronomer Galileo Galilei discovered the moons of
    a. Jupiter
    b. Saturn
    c. Mars
    d. Uranus

Fill-In Questions:

8. In science, all hypotheses and theories must be tested by using them to make _____.
9. The variety of living organisms in a specific area is often referred to as _____.
10. The elements of an experiment that can be manipulated by a scientist are called _____ variables.
11. The results of an experiment that can be attributed to the manipulations of the experiment are called _____ variables.
12. The Periodic Table of the Elements was developed by _____.
13. The studies of creationism, astrology, and extrasensory perception (ESP) are classified as

    _____.
14. Science that is practiced without leading to practical applications is usually referred to as _____ research.
15. The application of the results of science to specific commercial or industrial goals is referred to as

    _____.

Crossword Quiz: Science: A Way of Knowing

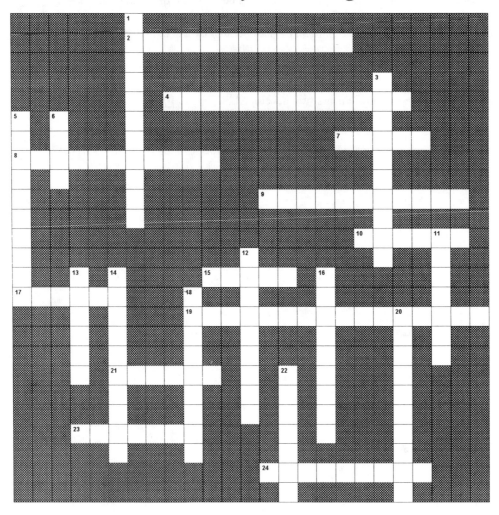

ACROSS

2. A term that refers to the number of different species in a certain area.
4. Ideas erroneously thought to be science
7. A hypothesis
8. Forecasts based on hypotheses and theories.
9. All good scientific theories must be _____.
10. An ecologist who studied nitrogen's effect on plant growth
15. The type of research that is knowledge for its own sake
17. A conclusion based on scientific observations or experiments.
19. SETI is the search for _____ intelligence
21. English physician who studied the circulation of blood
23. _____ is a way of asking and answering questions about the physical universe.
24. The supposed influence of the planets and stars on human life.

DOWN

1. The accurate notation or record of an event.
3. The _____ method is used to discover new scientific facts.
5. An activity sed to test a hypothesis.
6. This type of embryonic cell has the ability to form into any kind of mature cell.
11. The type of science that relates to practical systems.
12. The scientific study of stars and planets.
13. The band of background stars through the sun seems to move.
14. A tentative guess at the solution to a problem.
16. A prediction of daily events based on the positions of the planets and stars.
18. Russian chemist who proposed the Periodic Table of the Elements
20. Application of the results of science to commercial goals.
22. The amount of living material within a specified area.

Answers to Review Questions

Multiple-Choice Questions

1. b; 2. d; 3. d; 4. d; 5. c; 6. b; 7. a

Fill-In Questions

8. predictions; 9. biodiversity; 10. independent; 11. dependent; 12. Mendeleev; 13. pseudoscience; 14. basic; 15. applied science

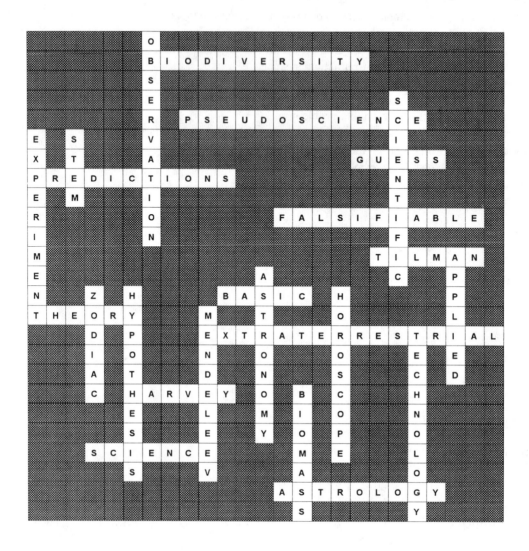

Chapter 2

The Ordered Universe

Chapter Review

After several centuries of careful observations, scientists have discovered that the behavior of objects, both enormous and small, in the universe can be described and predicted by a series of rules or principles often referred to as "laws." These laws describe the motion or velocity of objects based on the amount of material, or mass, an object contains, and the force required to move that mass. Using these laws, scientists have described such concepts as gravity, weight, and momentum. These laws also make it possible for scientists to predict the trajectories of the space shuttle and the placement of earth satellites used to transmit telephone conversations and television programs as they orbit the earth.

Learning Objectives

After studying this chapter, you should be able to:
(Other objectives may also be assigned by your instructor.)

1. Describe how the monument of Stonehenge was used by its builders to mark the passage of the year.
2. Relate how the process of scientific investigation studied in Chapter 1 was important in the discovery of the relationship between the disease cholera and human waste.
3. Compare the models of planetary motion proposed by the early astronomers Ptolemy and Copernicus.
4. Describe how Johannes Kepler used the earlier observations of Tycho Brahe to more accurately describe planetary orbits.
5. Discuss the historical importance of the scientist Galileo Galilei, and his use of technology (the telescope and the inclined plane) to advance scientific knowledge.
6. Differentiate between mass and weight.
7. Distinguish between uniform velocity and acceleration, and calculate these values for an object relative to its mass.
8. Define the gravitational force and show how it is related to the mass of two bodies and the distance between them.
9. Calculate what you would weigh on the surface of the moon.

Key Chapter Concepts

- In the ordinary world, we expect to discover natural causes for observed events.
- Predictability is the central principle of science.

- Observations over time allow us to make predictions about the positions of stars and planets during the seasonal changes in the night sky.
- Stonehenge, a large circular monument of stones built more than 4,000 years ago, served as a giant astronomical calendar used to mark the passage of the seasons.
- Scientists tend to accept the simplest explanation for a phenomenon as the most likely explanation.
- Ancient scholars believed that the universe was perfect and that all movements of stars and planets were based on spherical models.
- Spherical models did not explain the retrograde (apparent backward) motion of some planets.
- Subsequent study revealed that the planetary orbits were actually elliptical, not spherical.
- Proper analysis and interpretation of observations in the universe depend on the development of precise instruments and the application of mathematical techniques.
- Mechanics is the branch of science that deals with the motions of objects.
- Gravity is the attractive force between any two bodies in the universe.
  - The law of universal acceleration was originally proposed by Newton, and has been overwhelmingly confirmed by scientific observations.
  - Excessive acceleration, such as those experienced by astronauts and jet pilots may cause severe physiological effects
- Three universal laws of motion act in the universe:
  - The First Law states that an object at rest will remain at rest, or a moving object will continue moving in a straight line, until acted on by an unbalanced force. The tendency for an object to remain at rest or in uniform motion is called is called inertia.
  - The Second Law states that the acceleration produced on a body by a force is proportion to the magnitude of the force and inversely proportional to the mass of the object.
    - The second law tells us that the harder you throw a ball, the faster (and therefore, farther) it goes; and the more massive the ball, the shorter the distance it can be thrown.
  - The Third Law states that for every action there is an equal and opposite reaction.
    - When one object, a baseball bat for example, pushes on another, such as a baseball, the ball also pushes against the bat, altering its motion.
    - Forces always act simultaneously in pairs.
    - Every motion in life, from the swimming of a fish to the flight of a rocket ship, involves the interplay of all three of Newton's laws.
- Weight is the measure of the force of gravity on an object located at a particular point, and differs depending on where the object is located.
- Mass, the amount of matter present in an object, stays the same no matter where the object is located.
- Even though the universal laws of motion explain all movement in the universe, the uncertainty of actual events precludes our ability to accurately predict events with absolute certainty.

Key Individuals

- John Snow persuaded European cities to separate human waste disposal sites from drinking supplies and helped to reduce the occurrence of cholera.
- William Occam, an early English philosopher proposed that, given a choice, the simplest solution to a problem is most likely to be the correct answer.
- In the 1890's, Robert Koch discovered that cholera was caused by bacteria.
- Claudius Ptolemy proposed a system with the earth at the center of the universe (geocentric), and the planets' orbits as spheres-within-spheres.
- Nicolas Copernicus modified the Ptolemian hypothesis by placing the sun at the center

(heliocentric) of the solar system.

- Tycho Brahe accumulated extensive precise measurements of stars and planets.
- Johannes Kepler used Brahe's data to show that the planetary orbits were actually elliptical, not spherical.
- Galileo Galilei was the first scientist to observe heavenly bodies with a telescope. He also conducted a series of classical experiments analyzing the processes of speed, velocity, and acceleration.
- Galileo also discovered that no matter how heavy or light, any object dropped near the earth's surface accelerated at the same constant rate.
- Isaac Newton described the three universal laws of motion. These laws are still used to describe the motion of all bodies in the universe.
- Edmond Halley used Newton's laws of motion to calculate the orbit of the comet that bears his name, and he accurately predicted its return.
- Henry Cavendish determined the universal gravitational constant.

Key Formulas and Equations

- Velocity (in m/s) = distance traveled (in m)/time of travel (in sec), or $v = d/t$
- Velocity includes information on the direction of travel.
- Acceleration is change in velocity divided by the time it takes that change to occur: acceleration (m/s^2) = [final velocity - initial velocity] (m/s) / t (s), or $a = (v_f - v_i)/t$. Acceleration may be positive (speeding up) or negative (slowing down).
- When velocity changes, it is in certain units per second each second, described as meters per second per second (usually abbreviated m/s^2).
- The velocity of an accelerating object that starts from rest is proportional to the length of time that it has been falling: velocity (m/s) = constant a (m/s^2) x time (s), or $v = a$ x t
- The acceleration toward the earth is called gravity (g) and equals 9.8 m/s^2, or 32 $feet/s^2$.
- The distance covered by an accelerating object depends on the square of the travel time: distance traveled (m) = ½ x acceleration (m/s^2) x time^2 (t^2), or $d = ½ at^2$
- In equation form, the second universal law of motion may be stated: force = mass (in kg) x acceleration (in m/s^2), or $F = m$ x a. The unit of force is expressed in Newtons (N) as "kilogram-meter-per-second-squared" $(kg-m/s^2)$.
- Momentum is the product of an object's mass times its velocity: momentum (kg-m/s) = mass (kg) x velocity (m/s), or $p = mv$
- Newtons universal law: force of gravity (in Newtons [N]) = [G x mass$_1$ (in kg) x mass$_2$ (in kg)]/ [distance (in m)]^2, where G is a number known as the gravitational constant. The universal gravitational constant is 6.67 x 10^{-11} m^3/sec^2-kg, or 6.67 x 10^{-11} N-m^2/kg^2

KEY CONCEPT: Science Advances with Accurate Observations

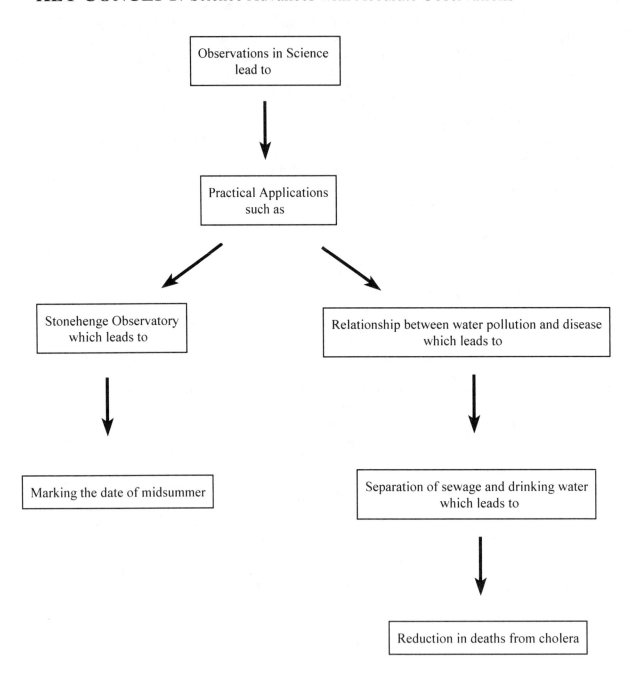

Observations in Science lead to

↓

Practical Applications such as

Stonehenge Observatory which leads to

Relationship between water pollution and disease which leads to

Marking the date of midsummer

Separation of sewage and drinking water which leads to

Reduction in deaths from cholera

Key Concept: Historical Development of Modern Astronomy

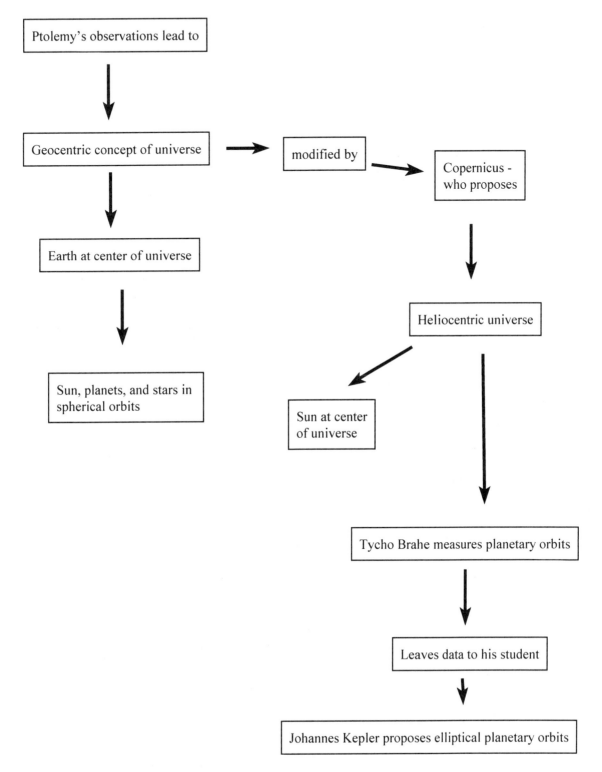

Ptolemy's observations lead to

Geocentric concept of universe

modified by

Copernicus - who proposes

Earth at center of universe

Heliocentric universe

Sun, planets, and stars in spherical orbits

Sun at center of universe

Tycho Brahe measures planetary orbits

Leaves data to his student

Johannes Kepler proposes elliptical planetary orbits

Key Concept: Mechanics is the Science that Studies Motion

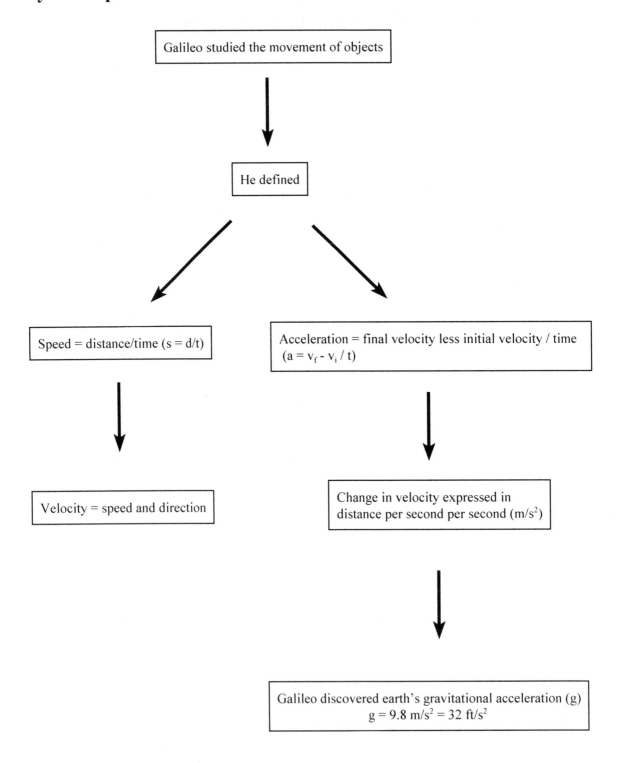

Galileo studied the movement of objects

He defined

Speed = distance/time (s = d/t)

Acceleration = final velocity less initial velocity / time $(a = v_f - v_i / t)$

Velocity = speed and direction

Change in velocity expressed in distance per second per second (m/s^2)

Galileo discovered earth's gravitational acceleration (g)
$g = 9.8 \ m/s^2 = 32 \ ft/s^2$

Key Concept: Isaac Newton's Laws of Motion

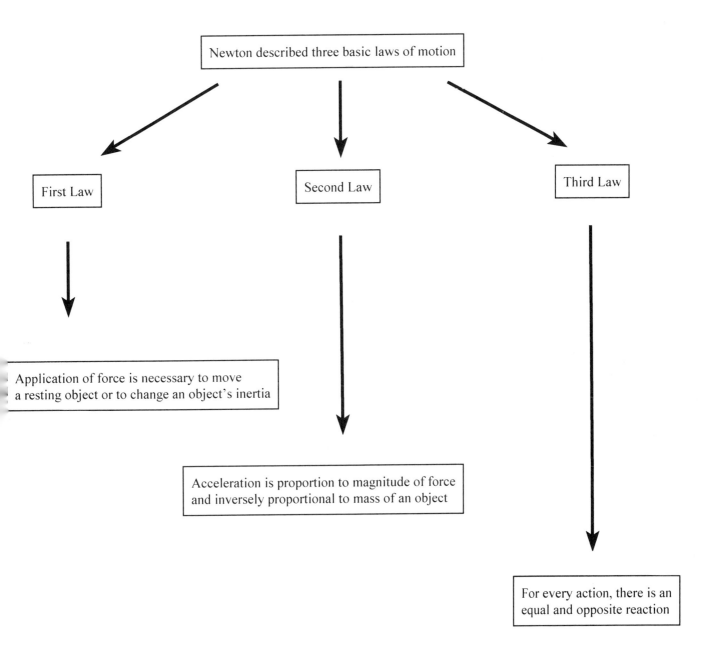

Newton described three basic laws of motion

First Law

Second Law

Third Law

Application of force is necessary to move a resting object or to change an object's inertia

Acceleration is proportion to magnitude of force and inversely proportional to mass of an object

For every action, there is an equal and opposite reaction

Self-Test

Multiple-Choice Questions:

1. Because of its practical applications to everyday life, one of the first sciences to develop was
   a. biology
   b. astronomy
   c. chemistry
   d. mechanics

2. Stonehenge appears to have been used by ancient groups of people to
   a. predict the phases of the moon
   b. mark the beginning of winter
   c. mark the beginning of summer
   d. mark the middle of summer

3. When confronted with a problem in the physical world, scientists usually
   a. accept the simplest solution most likely to be right
   b. look for the most complicated solution to the problem
   c. develop new mathematical models to explain the problem
   d. wait for future generations to solve the problem

4. The Danish astronomer who first used a sloping device called a "quadrant" to observe and record the positions of the planets and stars was
   a. Copernicus
   b. Ptolemy
   c. Brahe
   d. Kepler

5. The first astronomer to propose that the planets traveled in elliptical, not circular, orbits was
   a. Copernicus
   b. Ptolemy
   c. Kepler
   d. Galileo

6. The first astronomer to observe the planets with a telescope was
   a. Copernicus
   b. Ptolemy
   c. Kepler
   d. Galileo

7. Which of the following individuals was imprisoned because he popularized the idea that the sun, and not the earth, was the center of the solar system?
   a. Newton
   b. Copernicus
   c. Kepler
   d. Galileo

8. Which of the following individuals utilized an inclined plane to study the movement falling objects?
    a. Newton
    b. Copernicus
    c. Kepler
    d. Galileo

9. The expression of "velocity" differs from the expression of "speed" because
    a. velocity is always faster than speed
    b. speed is always faster than velocity
    c. velocity always indicates the direction of travel
    d. speed always indicates the direction of travel

10. "For every action there is an equal and opposite reaction" is
    a. Newton's first law of motion
    b. Newton's second law of motion
    c. Newton's third law of motion
    d. none of the above

11. "The acceleration produced on a body by a force is proportional to the magnitude of the force and inversely proportional to the mass of the object" is
    a. Newton's first law of motion
    b. Newton's second law of motion
    c. Newton's third law of motion
    d. none of the above

12. "A moving object will continue moving in a straight line at a constant speed, and a stationary object will remain at rest, unless acted on by an unbalanced force" is
    a. Newton's first law of motion
    b. Newton's second law of motion
    c. Newton's third law of motion
    d. none of the above

Fill- Questions:

13. _____ was the Greek philosopher who devised an explanation of planetary movement involving "spheres-within-spheres."
14. A heliocentric view of the universe postulates that the _____ is at the center of the solar system.
15. _____ was the European philosopher who first analyzed Tycho Brahe's data mathematically.
16. A geocentric view of the universe postulates that the _____ is at the center of the solar system.
17. _____ invented the first thermometer and the pendulum clock.
18. Any change in velocity is described as _____.
19. Velocity always has the same numerical value as _____.
20. The units of acceleration are _____.

Problems:

1. Albert weighs 150 pounds on earth.
   a. How much does Albert weigh in newtons?
   b. What is Albert's mass in grams? In kilograms?

2. Gravity on the moon is approximately one-sixth (about 0.17) the gravity on earth.
   a. How much would Albert weigh on the moon?
   b. What is Albert's mass on the moon in grams? In kilograms?

3. Albert is chosen to participate in a flight of the space shuttle, and once in orbit, he discovers that he is weightless.
   a. What is Albert's mass in orbit in grams? In kilograms?

4. You are at a baseball game, and the pitcher throws a fast ball that accelerates to the batter at 50 m/s^2. Assume that the baseball has a mass of 0.15 kg. How much force (in newtons) must the batter apply to the ball to lay down a perfect bunt that stops dead in front of home plate?

5. Assume that you are driving an automobile that has a mass of 800 kg. How much force (in newtons) must you apply to bring that auto from a dead stop to a speed of 50 m/s in 5 seconds?

Crossword Quiz: The Ordered Universe

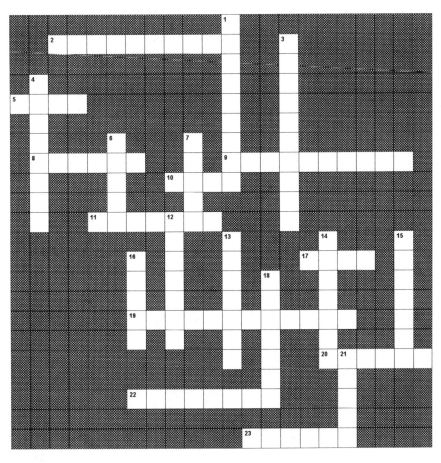

ACROSS

2. The apparent reversal of movement of a planet around the Sun
5. German scientist who linked cholera to a specific microorganism
8. He developed three fundamental laws of motion
9. He proposed that the Sun was the center of the universe
10. The amount of material present in an object
11. He proposed elliptical orbits for planets
17. London physician who showed that the spread of disease was linked to contamination of drinking water
19. Change in velocity per unit time
20. The movement of an object from one place to another
22. A measure of distance/time and direction
23. A measure of the force of gravity on an object

DOWN

1. The study of motion of material objects
3. An ancient observatory

4. This measure equals the product of an object's mass times its velocity
6. A push or pull that causes a change in acceleration of an object
7. He developed a "quadrant" to observe stars and planets
12. He proposed the notion of the Divine Calculator
13. Greek astronomer who proposed that Earth was the center of the universe
14. Motion in a straight line at constant speed
15. The first scientist to observe planets with a telescope
16. William of _____ argued that we should always accept the simplest answer to a question
18. The attractive force between two heavenly bodies
21. The path that a planet travels around the Sun

Answers to Review Questions

Multiple-Choice Questions

1. b; 2. d; 3. a; 4. c; 5. c; 6. d; 7. d; 8. d; 9. c; 10. c; 11. b; 12. a

Fill-In Questions

13. Ptolemy; 14. sun; 15. Kepler; 16. Earth; 17. Galileo; 18. acceleration; 19. speed; 20. meter/sec^2

Problems

1a. 667.2 newtons; 1b. 67,500 g; 67.5 kg
2a. 25.5 lb; 2b. 67,500 g; 67.5 kg
3. 67,500 g; 67.5 kg
4. 7.5 newtons
5. 8,000 newtons

Chapter 3

Energy

Chapter Review

Energy is defined as the ability to do work, and it exists in several different forms. Potential energy may be stored in many ways, including the electrical charge in a battery, the gasoline in an automobile's fuel tank, and the natural gas that feeds the burners in a furnace. Potential energy may be converted into other forms of energy, such as the light energy that a flashlight emits, the kinetic energy in the moving automobile, and the heat energy used to cook food. The work accomplished by energy is measured in joules (or foot-pounds), and defined as the use of a force through a distance. The rate at which energy is used is measured in watts or kilowatts. The first law of thermodynamics states that, even though energy can be converted from one form to another, the total amount of energy in a closed system remains constant. The earth is not a closed system. It constantly receives new energy from the sun, and loses energy to outer space.

Learning Objectives

After studying this chapter, you should be able to:
(Other objectives may also be assigned by your instructor.)

1. Define energy and indicate the different forms in which energy exists.
2. Discuss how one form of energy may be converted into another form.
3. Indicate how the concepts of work and power relate to energy.
4. Define the units (such as joule, watt, kilowatt) associated with describing the attributes of using energy.
5. Describe the phenomenon of wave energy.
6. Discuss the relationship between mass and energy.
7. Describe the energy balance of the earth.
8. Describe how the first law of thermodynamics and energy equations relate to the energy balance (and the weight gain or loss) in the human body.
9. Use standard mathematical equations to calculate the relationships among mass, energy, work, and power.
10. Relate the universal law of gravitation that you studied in Chapter 2 to gravitational potential energy studied in this chapter.

Key Chapter Concepts

- Energy is the ability to do work.
- Energy appears in a wide variety of forms:
  - Potential energy is energy that is stored and is capable of doing work when released.
    - Gravitational potential energy may be stored in an object such as a piledriver weight that is raised and dropped on another object.
    - Chemical potential energy is stored in fuels such as foods, gasoline, and batteries.
    - Electrical potential energy exists in the electrical wires in your home.
    - Elastic potential energy exists in a coiled spring or drawn bowstring.
    - Magnetic potential energy exists between two oppositely charged magnets.
  - Energy may be manifested in many different ways:
    - Kinetic energy is the energy of movement.
    - Radiant energy is the energy associated with light.
    - Heat or thermal energy is the kinetic energy of atoms and molecules.
    - Chemical energy is stored in the bonds between atoms.
    - Nuclear energy is bound within the nucleus of an atom.
  - Work is defined as the application of energy over a distance.
  - Power is defined as the amount of work done and the time it takes to do it.
  - Scientists first believed that heat was a fluid that flowed from place to place. Scientific experiments eventually showed that heat was actually a form of kinetic energy.
  - Wave energy, such as sound waves, are also a form of kinetic energy, the movement of molecules in another substance such as the air, water or a solid.
- Mass and energy are interconvertible.
  - One form of energy, such as electromagnetic or chemical energy, may be converted into another form of energy, such as heat or kinetic energy.
  - The First Law of Thermodynamics states that, within a closed system, even though energy may be converted from one form to another, the total amount of energy in the system remains constant.
  - In open systems, such as the earth, new energy from the sun constantly replaces energy that is lost to outer space.
  - Heat energy is measured in units called calories, the amount of heat needed to raise the temperature of a kilogram of water one degree C.
  - 3500 calories of energy taken in as food is converted into one pound of fat in the human body.
  - Energy is constantly flowing through the earth and all other systems of the universe.
  - Fossil fuels, such as coal, petroleum products, and natural gas, are classified as *nonrenewable* resources because they cannot be replaced within our lifetimes.

Key Individuals

- James Watt: defined units of work in terms of *horsepower*, the amount of work an average healthy horse could do every second.
- Benjamin Thompson: suggested that heat energy was a consequence of friction, and not an invisible fluid.
- Sir Humphry Davy: generated heat by rubbing two pieces of ice together on a cold London day.
- James Prescott Joule: showed that heat is another form of kinetic energy.
- William Thomson, Lord Kelvin: estimated the age of the Earth at less than 100 million years,

based on his calculations of heat lost from the surface of the planet. He was unaware of the presence of radioactive minerals in the interior of the earth that provide half of the earth's internal heat.

Key Formulas and Equations

- Work (in joules) = force (in newtons) x distance (in meters), or $W = F \times d$
- Power (in watts) = work (in joules) / time (in seconds), or $P = W / t$
- 1 watt of power = 1 joule of energy / 1 second
- Energy (in joules) = power (in watts) x time (in seconds)
- Kinetic energy (in joules) = ½ x [mass (in kg)] x [velocity (in m/s)]^2, or $E = \frac{1}{2} mv^2$
- Gravitational potential energy (in joules) = mass (in kg) x g (in m/s^2) x height (in m), or $E = mgh$
- Energy (in joules) = mass (in kg) x [speed of light (in m/s)]^2

Key Concept: Potential Energy is Expended in Several Ways

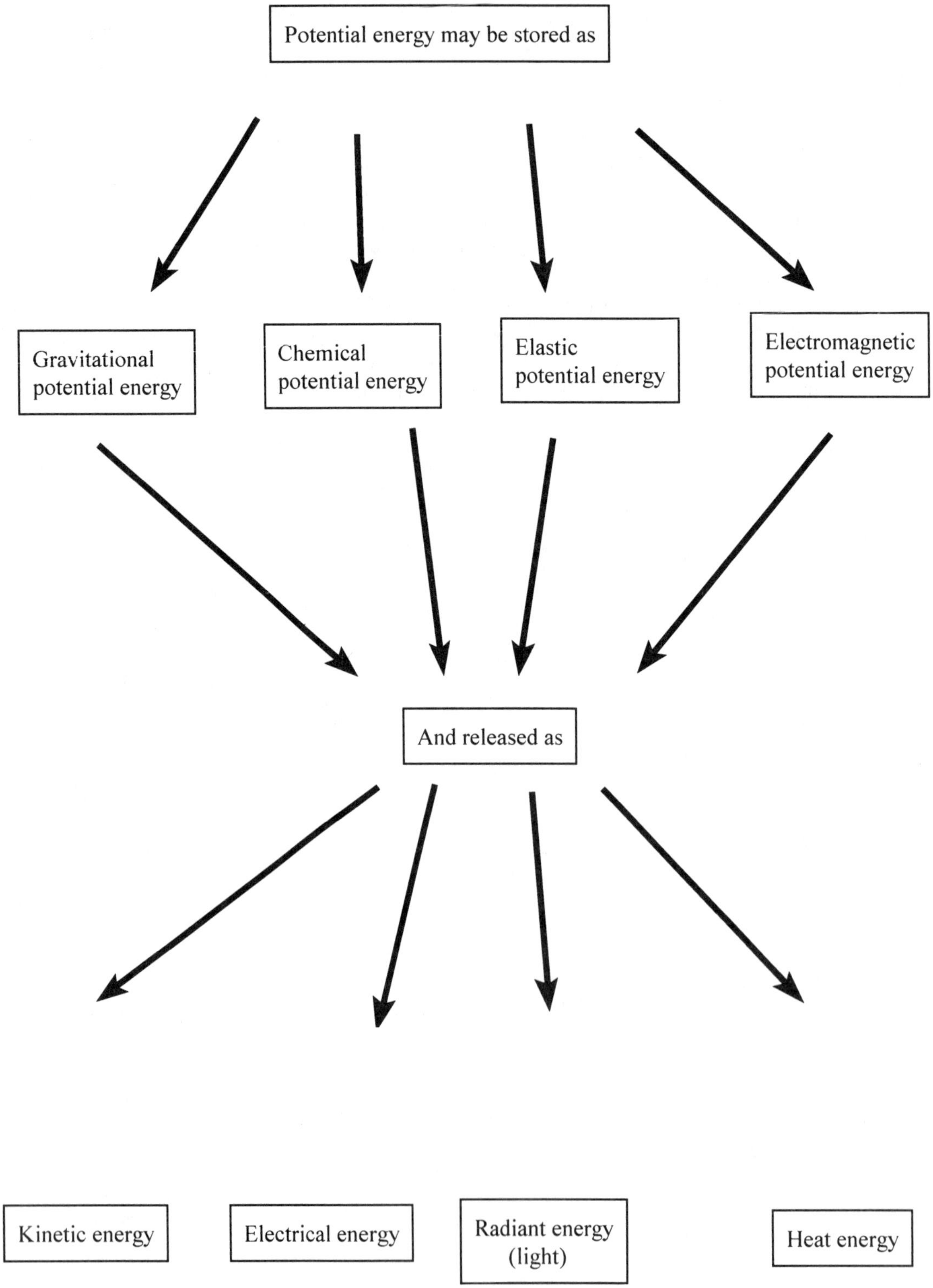

Potential energy may be stored as

Gravitational potential energy

Chemical potential energy

Elastic potential energy

Electromagnetic potential energy

And released as

Kinetic energy

Electrical energy

Radiant energy (light)

Heat energy

Key Concept: Energy Flows from the Sun Through Terrestrial Systems

Nuclear fusion in the sun produces light energy

↓

Light energy is converted to chemical energy by plants on earth

↓

Plants are buried and converted to coal and petroleum

↓

Miners and drillers bring coal and petroleum to surface

↙　　　　　　　　↘

Coal burned to produce steam which turns electric generators		Petroleum used to produce gasoline

↓　　　　　　　　↓

Electric energy causes light bulb to glow		Combustion of gasoline moves automobile

↘　　　　　　　　↙

All energy is eventually radiated back into space

Questions for Review

Multiple-Choice Questions:

1. The first law of thermodynamics states that
   a. more energy always leaves a system than comes into the system
   b. the total amount of energy in a closed system is conserved
   c. energy is equal to mass
   d. chemical potential energy can be converted only into kinetic energy

2. In the equation $E = mc^2$, the letter "c" stands for
   a. the chemical energy in the system under study
   b. the potential energy in the system under study
   c. the speed of light
   d. the kinetic energy in the system under study

3. Eventually, all energy generated on the earth is returned to space as
   a. heat
   b. kinetic energy
   c. work
   d. potential energy

4. In the equation $E = mc^2$, the letter "m" stands for
   a. meters
   b. molecules
   c. mass
   d. movement

5. The energy associated with a bowling ball knocking down bowling pins is
   a. kinetic energy
   b. potential energy
   c. chemical energy
   d. gravitational energy

6. Energy is expressed in terms of
   a. kilowatts
   b. newtons
   c. joules
   d. horsepower

7. When a force is exerted over a distance, we say that _____ has occurred.
   a. energy
   b. work
   c. power
   d. potential energy

8.   Work expended over a defined period of time is defined as
    a.   energy
    b.   newtons
    c.   force
    d.   power

9.   One joule exerted over one second is defined as
    a.   one watt
    b.   one newton
    c.   one kilogram
    d.   one calorie

10.   William Thomson, Lord Kelvin estimated the age of the earth to be
    a.   about 4 billion years
    b.   about 10 billion years
    c.   about 10 million years
    d.   about 100 million years

Fill-In Questions:

11.   The amount of energy needed to raise the temperature of a kilogram of water 1 degree C is defined as a _____.
12.   The major source of energy that powers activities on the earth is _____.
13.   The fluidized bed combustor is an example of a high-tech _____.
14.   Coal, petroleum, and natural gas are all examples of _____ fuels.
15.   Work equals force times the _____ over which it is exerted.
16.   A _____ is the amount of work done when you exert a force of one newton through a distance of one meter.
17.   _____ is the amount of work done, divided by the time it takes to do it.
18.   Energy is expressed in _____.
19.   Power is expressed in_____.
20.   Einstein's equation, $E = mc^2$, states that mass can be converted into _____ .

Problems:

1.   Let us assume that you are playing left field in a baseball game and you catch a fly ball hit out to you by a batter. How much work does it take for you to throw the ball back to the pitching mound? Assume that the baseball weighs 0.5 kg, and you are 50 meters from the pitching mound. (Hint: you are overcoming gravity which is $9.8 m/s^2$.)

2.   If it took the baseball in the previous problem 3 seconds to travel from left field to the pitcher, how much power was expended?

3.   Suppose that the electric company in your town charges $0.10 per kilowatt-jour for the electricity that you use. How much will the total cost be to you to burn a 100 watt light bulb for six hours a day for a full month (30 days)?

day for a full month (30 days)?

4. A massive object, such as an automobile, can cause much damage when it collides with another object, even though it may be traveling at a relatively slow speed. How much kinetic energy is released (in joules) when a small sedan (800-kg) collides with a lamp post at 5 meters per second (about 12 miles per hour)? How much energy is released if the speed is doubled?

5. According to Einstein's equation, how much potential energy is contained in the mass of a 2-gram cube of sugar? Approximately how many kilowatt-hours of power would this translate into?

Crossword Quiz: Energy

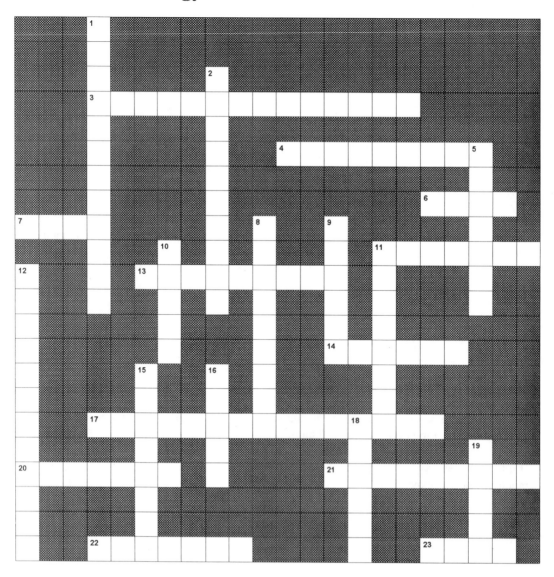

ACROSS:

3. The first law of _____ states that the total amount of energy in a closed system remains the same
4. Any animal that survives by eating other animals
6. The kinetic energy of atoms and molecules
7. Defined as one joule of energy expended over one second of time
11. The energy of movement
13. Energy that is stored in a system
14. The unit of measurement of force
17. Light is an example of this type of energy
20. A _____ level consists of all of the organisms that get their energy from the same source
21. Any animal that survives by eating plants
22. The amount of heat necessary to raise the temperature of one gram of water 1 degree Celsius
23. The application of a force through a distance

DOWN:

1. This type of material can convert sunlight energy into electrical energy
2. The English unit for measuring power
5. The type of potential energy found in a stretched rubber band
8. A fossil fuel
9. He used heat loss to calculate the age of the earth
10. The unit used to express work
11. A thousand watts
12. The first law of thermodynamics deals with this cocept of energy
15. The type of energy that holds atoms and molecules together
16. Defined as mass times acceleration
18. Defined as the ability to do work
19. Defined as the rate at which work is done

Answers to Review Questions

Multiple-Choice Questions:

1. b, 2. c, 3. a, 4. c, 5. a, 6. c, 7. b, 8. d, 9. a, 10. d.

Fill-In Questions:

11. calorie, 12. the sun, 13. furnace, 14. fossil, 15. distance, 16. joule, 17. power, 18. joules, 19. watts, 20. energy

Problems:

1. 245 joules
2. 81.6 watts
3. $1.80
4. 10,000 joules at 12 miles per hour, 40,000 joules at 24 miles per hour
5. 1.8×10^{14} joules, which translates to approximately 50 million kilowatt-hours

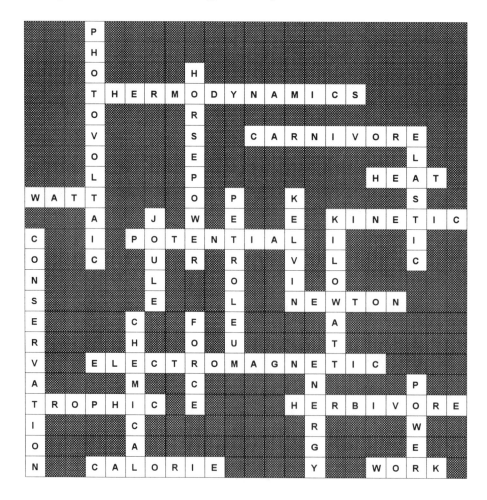

Chapter 4

Heat and the Second Law of Thermodynamics

Chapter Review

Heat (thermal energy) is a measure of the atomic kinetic energy contained in an object. The thermal energy present in any object in the universe is expressed as its temperature. Absolute zero is a temperature defined as the total absence of heat. All objects in the universe are at a temperature above absolute zero, and thus contain some thermal energy. The amount of heat that an object can contain is its heat capacity. Heat spontaneously flows from a warmer to a cooler object, and may be transferred through direct contact (conduction), by the movement of fluids or gases (convection), or by electromagnetic wave energy (radiation). The second law of thermodynamics states that the conversion of one form of energy to another is never totally efficient. When energy is converted, some is lost as heat, the least organized form of energy. All ordered systems tend to become disorganized over time, unless energy is added to the system.

Learning Objectives

After studying this chapter, you should be able to:
 (Other objectives may also be assigned by your instructor)

1.   Discuss the concept of time and directionality in nature.
2.   Define heat (thermal energy) and indicate how it fits into the discussion of energy in the previous chapter.
3.   Distinguish among the three scales (Fahrenheit, Celsius, and Kelvin) used to measure temperature, and convert from Fahrenheit to Celsius and from Celsius to Fahrenheit.
4.   Define the term *heat capacity*.
5.   Identify the three methods in which heat can be transferred from one object to another.
6.   Indicate how insulation serves to conserve heat in both nonliving and living systems
7.   Discuss the second law of thermodynamics, and indicate how it restricts the way heat and other forms of energy can be transferred.
8.   Describe the concept of efficiency in terms of the transfer of energy.

Key Chapter Concepts

- In nature, events tend to proceed from improbable situations to more probable situations.
    - Over a period of time, the atoms in any closed system tend to distribute themselves such that all have approximately the same energy and motion.
    - Heat (thermal energy) is a measure of the atomic kinetic energy contained in any object.

- Heat is often measured in *calories*, defined as the amount of heat required to raise the temperature of one gram of water one degree Celsius in temperature.
- Temperature, a measure of the heat energy present in an object may be expressed in three different scales: Fahrenheit, Celsius, and Kelvin.
- Heat capacity is a measure of the ability of a material to absorb heat energy. The heat capacity of a material is defined as the amount of heat required to raise the temperature of one gram of that material by 1 °C.
  - The heat capacity of water is one calorie.
- Heat energy is spontaneously transferred from one place or body to another.
  - Conduction is the movement of heat energy by collisions between vibrating atoms or molecules.
  - Convection is the transfer of heat by the physical motion of masses of gases or fluids.
  - Radiation is the transfer of heat by electromagnetic radiation.
  - Efficiency is the amount of work you get from a system (or engine) divided by the amount of energy you put into the system. Heat loss reduces efficiency in electrical and mechanical engines.
- Insulation retards the transfer of energy.
  - Wood, fiberglass, and masonry are good insulation materials used in houses to conserve the loss of heat.
  - Trapped air in double-pane windows is a good insulation material
  - Fur and feathers insulate mammals and birds, respectively, against excessive heat loss.
  - Clothing insulates the human body from excessive heat loss.
- The human body has several mechanisms to regulate internal temperature.
  - Blood vessels carrying warm blood near the surface of the body dilate to carry heat to the surface, where it can be radiated away, or constrict and conserve heat in the blood in the body's interior.
  - Sweating removes heat from the body by transferring heat to the water molecules that evaporate from the skin.
  - Shivering is involuntary muscular contraction that produces internal heat.
- The second law of thermodynamics states that the energy in every system will undergo spontaneous conversions from one form to another, but these conversions are inefficient, and some energy will be lost as heat to the surrounding environment.
  - Energy will not flow spontaneously from a cold body to a hot body.
  - Every isolated system will become progressively more disordered with time (entropy).
  - Every engine that exists must have an energy source (sunlight, fuel, or electricity).
  - The evolution of life on earth does not annul the second law. Living things have become progressively more complex because energy, in the form of sunlight is constantly added to the earth, and used by living things to run their engines.
- Aging is a practical example of the second law of thermodynamics in living things. As living things age, their body parts wear to the point of complete inefficiency, illustrating the principle of entropy.

Key Individuals In Science

- William Thompson, Lord Kelvin, who we introduced in Chapter 3, introduced the Kelvin temperature scale that defined *absolute zero*, the coldest temperature attainable anywhere in the universe.
- Charles Darwin's hypotheses about the evolution of living things appeared to annul the second law of thermodynamics. Later discoveries would show that evolution does not annul the second law.

Key Formulas and Equations

- The formula to convert degrees Celsius to degrees Fahrenheit is $°F = (1.8 \times °C) + 32$
- The formula to convert degrees Fahrenheit to degrees Celsius is $°C = (°F - 32) / 1/8$
- Efficiency is a comparison of the temperature *difference* between a high-temperature reservoir and a low-temperature reservoir divided by the high temperature reservoir:

    - efficiency (in %) = $\dfrac{(\text{temperature}_{hot} - \text{temperature}_{cold})}{\text{temperature}_{hot}} \times 100$

    - or, in symbols: $\text{Eff.} = \dfrac{T_{hot} - T_{cold}}{T_{hot}} \times 100$

Key Concept: Heat is Transferred in Several Ways

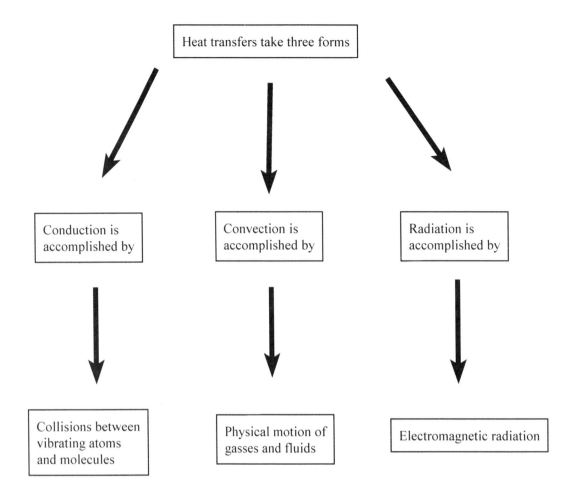

Questions for Review

Multiple-Choice Questions:

1. The measure of the quantity of heat contained in an object is referred to as its
    a. thermal energy
    b. heat capacity
    c. heat transfer
    d. temperature

2. The transfer of heat by direct atomic collisions is referred to as
    a. thermal conductivity
    b. conduction
    c. convection
    d. radiation

3. Water boils at $373°$ in the _____ temperature scale.
    a. English
    b. Kelvin
    c. Fahrenheit
    d. Celsius

4. Water freezes at $0°$ in the _____ temperature scale.
    a. English
    b. Kelvin
    c. Fahrenheit
    d. Celsius

5. The amount of heat energy required to raise the temperature of one gram of water by one degree Celsius is a
    a. thermal unit
    b. conductivity unit
    c. convection unit
    d. calorie

6. The transfer of heat by the bulk motion of gases or fluids is referred to as
    a. convection
    b. conduction
    c. radiation
    d. thermal movement

7. All isolated systems will spontaneously tend toward disorder. This phenomenon is referred to as
    a. thermal inefficiency
    b. heat transfer
    c. entropy
    d. thermal conductivity

8.   Which of the following statements is not consistent with the second law of thermodynamics?
a.   all isolated systems will tend to remain ordered indefinitely
b.   heat will not flow spontaneously from a cold body to a hot body
c.   no engine is one hundred percent efficient in converting energy to work
d.   the evolution of more complicated forms of life on earth does not annul the second law

9.   A fire transfers most of heat by
a.   convection
b.   conduction
c.   radiation
d.   infusion

10.   Which of the following materials has the highest thermal conductivity?
a.   wood
b.   metal
c.   air
d.   fur

Fill-In Question:

11.   The sun transfers most of its heat energy to the earth by means of _____.
12.   _____ is the temperature at which all atomic and molecular kinetic motion ceases.
13.   Water boils at _____° in the Fahrenheit temperature scale.
14.   _____ is a measure of the ability of a material to absorb heat energy.
15.   Ten calories of heat energy will raise the temperature of ten grams of water ____° C.
16.   _____ is the method used by a forced air heater to heat a house.
17.   _____ is the most commonly used insulation material in homes to prevent the loss of heat through the walls of the house.
18.   _____ is the amount of work you get from an engine divided by the amount of energy you put into it.
19.   Because they cannot be replaced in a reasonable time frame relative to human life, fossil fuels are often referred to a _____ sources of energy.
20.   _____ is a measure of the disorder in a physical system.

Crossword Quiz: Heat and the Second Law of Thermodynamics

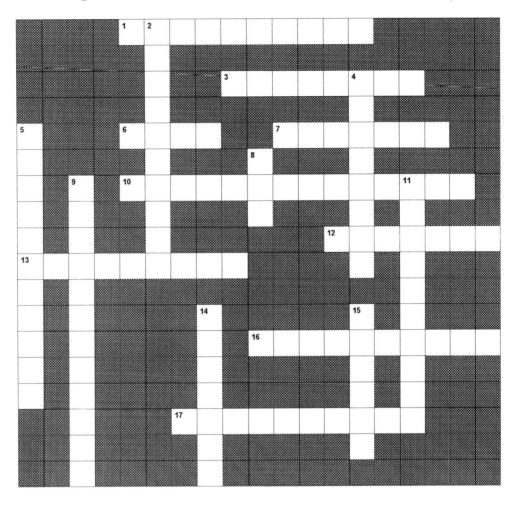

ACROSS

1. The amount of work performed by an engine divided by the amount of energy used
3. Specific heat _____ is the measure of the ability of a material to absorb heat energy
6. A form of energy that flows from a warmer object to a cooler object
7. The amount of heat needed to raise one gram of water one degree Celsius
10. The study of the movement of heat
12. A temperature scale in which the boiling point of water is designated as 100 degrees
13. The transfer of heat through electromagnetic activity
16. The transfer of heat through the mass movement of gases or fluids
17. The transfer of heat by direct atomic or molecular kinetic motion

DOWN

2. A temperature scale in which the boiling point of water is designated as 212 degrees
4. The entropy of an _____ system remains constant or increases
5. Measure of the warmth or coldness of an object
8. Heat will not flow spontaneously from a cold body to a _____ body
9. The measure of the ability of a material to transfer heat from one molecule to the next by conduction
11. Any material that inhibits the transfer of heat
14. The tendency toward disorder in a closed system
15. A temperature scale that includes "absolute zero"

Answers to Review Questions

Multiple-Choice Questions:

1. a; 2. b; 3. b; 4. d; 5. d; 6. a; 7. c; 8. a; 9. c; 10. b

Fill-In Questions:

11. radiation; 12. absolute zero; 13. 212°; 14. heat capacity; 15. 1°; 16. convection; 17. fiberglass; 18. efficiency; 19. nonrenewable; 20. entropy.

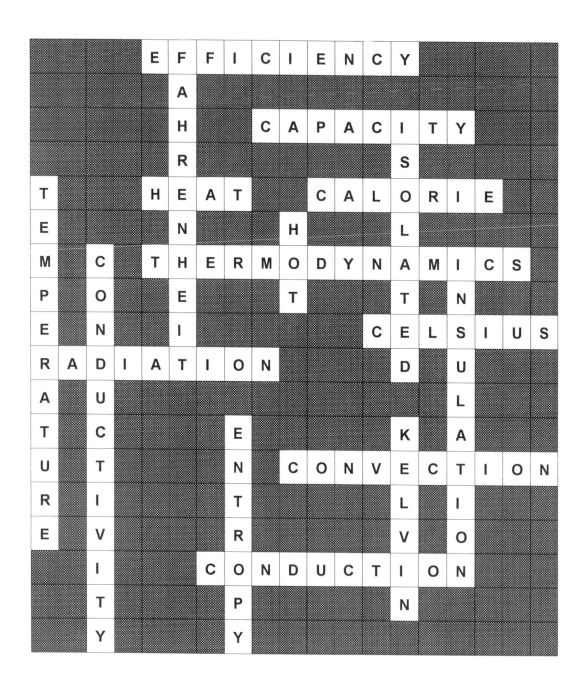

Chapter 5

Electricity and Magnetism

Chapter Review

Electricity and magnetism are different aspects of a major phenomenon present in the universe, the electromagnetic force. Electricity may take the form of static electricity or an electrical current. Static electricity is an electrical charge that is stationary, remaining with an object. Static electricity is due to a transfer of electrons between two objects. The object that accumulates excess electrons also accumulates a negative charge, while the object that loses electrons assumes a positive charge. The oppositely charged bodies attract one another. Electrical current is associated with the movement of electrons. Direct current (DC) involves the one-way movement of electrons, while alternating current (AC) involves a flow of electrons that reverses its direction at regular intervals. Characteristics associated with electricity are strength of current (measured in amperes), differences between currents (measured in voltage), and the resistance to the passage of an electrical current (measured in ohms). Magnets exhibit force fields similar to electrical fields, characterized by opposite (north and south) poles.

Learning Objectives

After studying this chapter, you should be able to:
(Other objectives may also be assigned by your instructor)

1. Describe the concept of an electrical charge and distinguish between static electricity and an electrical current.
2. Discuss how the movement of electrons can create positive and negative electrical charges.
3. Use Coulomb's law to calculate the force that exists between two electrically charged objects.
4. Compare and contrast an electrical field with a magnetic field.
5. Discuss some of the practical uses of magnets in modern technology.
6. Describe how a storage battery can accumulate an electrical charge.
7. Discuss the electromagnet and its commercial uses.
8. Describe how an electrical circuit forms an unbroken path that carries electricity.
9. Distinguish among amperage, voltage, and resistance in an electrical circuit.
10. Describe how electrical impulses are used in the human body to transmit information from one place to another.

Key Chapter Concepts

- The movement of electrons from one object to another produces an electrical charge.

- When electrons leave an object, unbalanced positive charges in the nucleus give the object a net positive electrical charge.
- When electrons accumulate on an object, the object acquires a net negative electrical charge.
- Once placed on an object, an electrical charge that does not move is called static electricity.
- Like charges repel one another, while unlike charges attract one another.
- Objects carrying a static electrical charge emit an electric field in all directions that becomes weaker at greater distances from the object.
- A magnet is a metallic object that attracts other metallic objects, such as iron.
  - Every magnet has at least two poles, and like magnetic poles repel each other, while unlike poles attract.
  - The poles of a magnet are usually labelled north and south.
  - Magnetic lines of force extend from the north to the south pole, creating a magnetic field (called a dipole field) surrounding the magnet.
  - The presence of iron ore within the earth results in the earth behaving like a giant magnet with north and south poles.
  - The attraction between opposite ends of magnets allows small magnetized pieces of iron to be used as compass needles; one end pointing toward Canada in the north, and the other end pointing toward Antarctica in the south.
- Magnetic and electrical forces share many features.
  - Magnetic fields can be created by the motion of electrical charges.
  - An electromagnet is a device consisting of a coil of wire that produces a magnetic field whenever an electrical charge runs through the wire.
  - Electromagnets are used in many devices, including switches, buzzers, and electric motors.
  - Electric motors employ rotating electromagnets within a permanent magnet. As the direction of flow of electrons in the electromagnet is alternated, the electromagnet spins, operating such devices as electric fans, electric hand drills, and the motors that raise and lower the windows in your car.
  - The electric current in one electromagnet can induce a current in a separate nearby electromagnet through electromagnetic induction.
- Electrical generators can be used to generate an electrical current.
  - If a loop of wire is spun between the north and south poles of a large magnet, an electric current is generated in the wire.
  - The mechanical energy used to spin the wire may be provided by water flowing through a dam, steam produced in a coal-burning furnace or nuclear reactor, or wind-driven propeller blades.
  - Energy provided by gasoline burned in an automobile engine turns a rotor in a generator (or alternator) and generates electricity to power the headlights, radio, and power windows, and is stored in the car's battery.
  - Because current produced in a generating coil flows first one way and then the other, it is called alternating current.
  - The current that flows from a storage battery flows in one direction only, and is called direct current.
- Electric circuits are closed loops of material that carry an electrical current.
  - Every circuit contains : a source of energy (a battery or generator), a closed metal wire path, and a device (such as a motor or light bulb) that uses the electrical energy.
  - Three qualities characterize every circuit: current, voltage, and resistance.
  - Current, measured in amperes (or amps), measures the amount of electric charge that passes

through the circuit. One amp = the flow of one coulomb (the unit of electric charge) per second past a point in the wire.

- Voltage, measured in volts, is a measure of the pressure produced by the energy source to push the current through the wire; the greater the voltage, the faster the current flows through the circuit.
- Electrical resistance, measured in ohms, measures the difficulty the current experiences in passing through the circuit. The higher the resistance, the more of the electrical energy is converted into heat, which reduces the efficiency of the system.
- In series circuits, two or more loads are linked along a single loop of wire. In parallel circuits, different loads are situated on different wire loops.
- Practically all cellular activities in the body involve the movement of an electrical current. The electrical signals generated by nerve cells are the best examples of electrical activity in the body.

Key Individuals in Science

- Benjamin Franklin (1706-17-90) was an early pioneer of electrical science. He performed potentially lethal experiments with kites in thunderstorms that showed the electrical nature of lightning. He also invented the lightning rod, a device that diverts the electric charge of a lightning bolt away from a building and into the ground.
- Charles Augustin de Coulomb (1736-1806) defined the nature of the electrical force between two charged bodies. Coulomb's law states that the force between any two electrically charged objects is proportional to the product of their charges divided by the square of the distance between them.
- Luigi Galvani (1737-1798) showed that an electrical jolt could induce convulsive twitching in amputated frogs' legs.
- Allesandro Volta (1745-1827) developed the first storage battery, a device that converted chemical energy in the battery materials into the kinetic energy of electrons running through an outside wire.
- Han Christian Oersted (1770-1851) discovered that a magnetic field could create the motion of electrons that resulted in an electrical charge. His work led to the development of the electromagnet.
- Michael Faraday (1791-1867) performed a series of classic experiments through which he discovered a central idea that helped link electricity and magnetism.
- James Clerk Maxwell (1831-1879) discovered four fundamental laws of electricity and magnetism:
  - Coulomb's law - like charges repel, unlike attract.
  - There are no magnetic monopoles in nature.
  - Magnetic phenomena can be produced by electric effects
  - Electrical phenomena can be produced by magnetic effects.

Key Formulas and Equations

- electrostatic force (in newtons) = k x $\frac{\text{1st charge x 2nd charge (in coulombs)}}{\text{distance}^2}$ or, $F = k \frac{q_1 \times q_2}{d^2}$

- 1 coulomb = the charge on 6.3×10^{18} electrons

- electric power (in watts) = current (in amps) x voltage (in volts)

Key Concept: Relationship Between Movement of Electrons and Electric

Charge

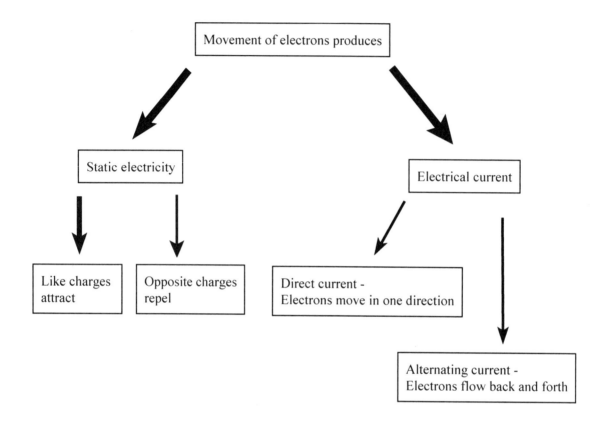

Key Concept: Connections Between Electrons and Magnetism

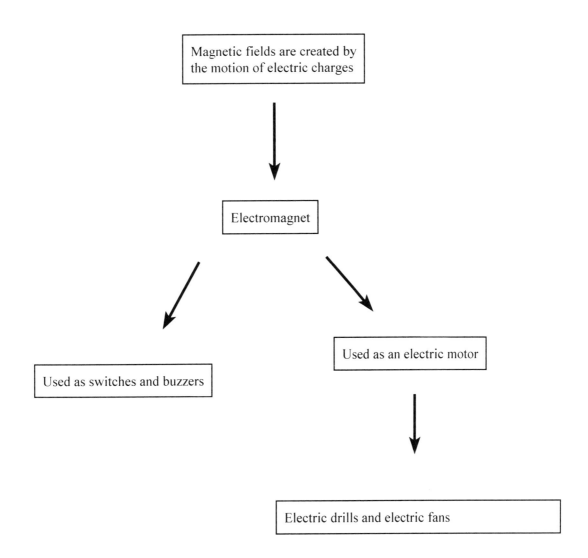

Key Concept: Electric Circuits

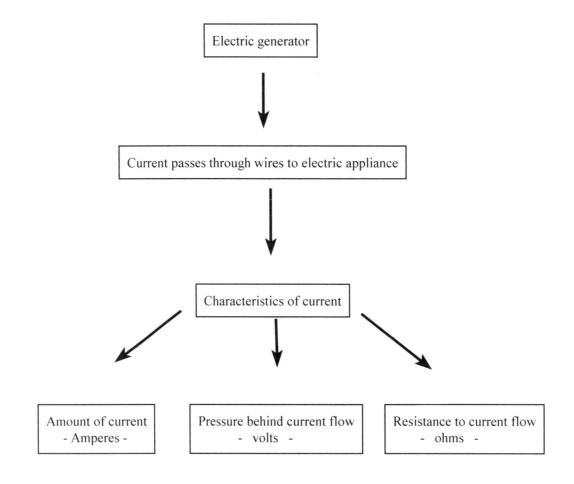

Questions for Review

Multiple-Choice Questions:

1. The force that moves electrically charged objects toward and away from each other is
    a.   electrostatic force
    b.   amperage
    c.   a magnetic monopole
    d.   voltage

2. Objects from which electrons are removed
    a.   gain a negative charge
    b.   gain a positive charge
    c.   gain both a positive and negative charge
    d.   gain no electric charge at all

3. The historical figure who demonstrated that lightning was electrical in nature was
    a.   Michael Faraday
    b.   Luigi Galvani
    c.   Allesandro Volta
    d.   Benjamin Franklin

4. The person who defined the nature of the electrical force between two charged bodies was
    a.   Michael Faraday
    b.   Charles Coulomb
    c.   Allesandro Volta
    d.   Benjamin Franklin

5. Coulomb's law states that the force of static electricity between two objects is inversely proportional to
    a.   the amount of charge on the objects
    b.   the square of the distance between the objects
    c.   the direction of the charge on the objects
    d.   the size of the objects

6. The collection of lines that map out the directions in which compass needles would point is called a(an)
    a.   electric field
    b.   magnetic dipole
    c.   magnetic field
    d.   compass field

7. A device that converts stored chemical energy into the kinetic energy of electrons passing through an outside wire is called
    a. a battery
    b. a compass
    c. an electromagnet
    d. a motor

8. The amount of electrons flowing in a specific direction is termed the
    a. voltage
    b. current
    c. electrical resistance
    d. electric circuit

9. The dipole field is found in
    a. bar magnets
    b. static electric balls
    c. lightning bolts
    d. electric motors

10. An electric current flowing from a storage battery is
    a. alternating current
    b. direct current
    c. electromagnetic current
    d. dipole current

Fill-In Questions:

11. Objects that have an excess of electrons are said to carry an _____.
12. The unit of electrical charge is the _____.
13. The collection of all the arrows that represent the forces around a charged object is known as the

    _____.
14. Like magnetic poles repel one another, while unlike poles _____.
15. The first storage battery was developed by _____.
16. Rotating an electromagnet inside a permanent magnet can be used to make an electric _____.
17. The closed pathway that carries an electrical current is called a _____.
18. The number of electrons in an electrical current are measured in _____.
19. The cell in the human body that is specialized for transmitting electrical impulses is the _____ cell.
20. A negatively charged object has an abundance of _____.

Problems:

1. Suppose you are using a flashlight to illuminate your way back to your tent in a campground after visiting the restroom. The battery in the flashlight is 12 volts and the bulb in the flashlight is 2 amps. How much power do you use to cross the short distance to your tent?

2. Suppose that your power company charges $1.00 per kilowatt-hour as its power rate. How much

2.   Suppose that your power company charges $1.00 per kilowatt-hour as its power rate. How much does it cost you to run your 5-amp air conditioner on your standard 115-volt circuit in your house for 12 hours?

3.   How much current does your 100-watt television set use when it is plugged in and turned on in your standard 115-volt electrical system at home? How many coulombs is this if you play the television for 5 hours?

4.   Which is more efficient for you to use at home: a 10-amp air conditioner for one hour to completely cool of your room, or a 3-amp electric fan to keep a constant flow of air over you for five hours?

Crossword Quiz: Electricity and Magnetism

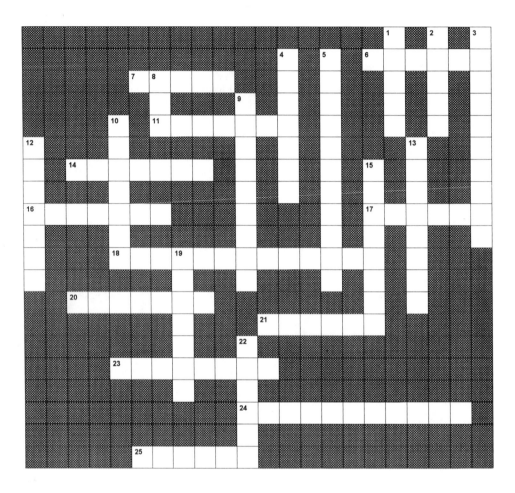

ACROSS

6. The magnetic field that arises from the two poles of a magnet
7. The rate at which work is done
11. Discovered four fundamental laws of electricity and magnetism
14. Used electricity to cause twitching of frog muscles
16. A needle-shaped magnet designed to point toward the poles of the Earth's magnetic field
17. An electric charge that does not move
18. A device that produces a magnetic field from a moving electric charge
20. Converts stored chemical energy into kinetic energy of charged particles
21. The unit used to measure the magnitude of current
23. He discovered the electrical nature of lightning
24. Energy made available by the flow of electric charge through a conductor
25. An object that exhibits a north and south pole

DOWN

1. The electric force that would be felt at a particular point around a charged object
2. A rotating machine that transforms electrical energy into mechanical energy
3. Anything that impedes the flow of an electric current
4. The unit for measuring the magnitude of an electric charge
5. A type of electric current in which charges reverse their direction of motion
8. A unit of measurement for the electrical resistance of a wire
9. Produces an alternating current in an electric curcuit through the use of electromagnetic induction
10. The pressure produced by the energy source in a circuit
12. A pathway through which an electric current flows
13. Any object with an oversupply of electrons has a _____ electric charge
15. Any object from which electrons have been removed has a _____ electric charge
19. The flow of electrons from one point to another
22. A type of electric current in which the electrons flow in one direction only

Answers to Review Questions

Multiple-Choice Questions

1. a; 2. b; 3. d; 4. b; 5. b; 6. c; 7. a; 8. b; 9. a; 10. b

Fill-In Questions

11. electric charge; 12. coulomb; 13. magnetic field; 14. attract; 15. Volta; 16. motor; 17. circuit; 18. amperes; 19. nerve; 20. electrons

Problems

1.   24 watts
2. $6.90
3.   0.87 amps
4.   1150 watts for air conditioner versus 1725 watts for electric fan

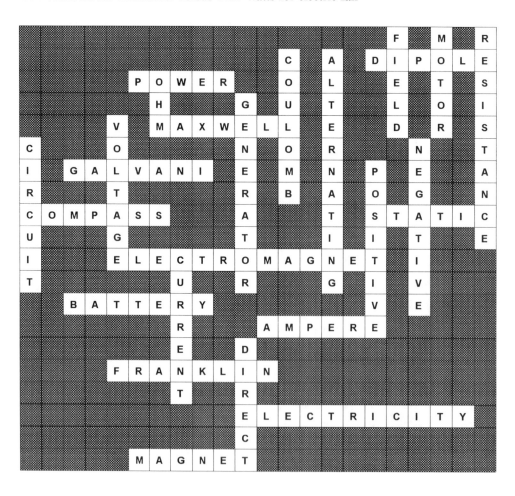

Chapter 6

Electromagnetic Radiation

Chapter Review

Electromagnetic radiation is a type of energy that is transmitted in the form of a wave. We can measure several properties of waves, including electromagnetic waves: wavelength, frequency, velocity, and amplitude. Two waves may enhance or decrease their combined activities through constructive or destructive interference. James Maxwell devised equations to describe the characteristics of electromagnetic waves. Different types of electromagnetic radiation, such as sound, light and ultrasonic waves, have great influence on the activities of living things. Electromagnetic radiation travels at the speed of light, a universal constant. The variety of electromagnetic radiation covers a wide spectrum, determined by the wave length and frequency of the particular type of radiation. The visible spectrum of light occupies a very narrow band in the total electromagnetic spectrum. The total spectrum varies from very long radio waves to extremely short X-rays and gamma rays.

Learning Objectives

After studying this chapter, you should be able to:
(Other objectives may also be assigned by your instructor)

1. Define a wave, and describe how it transfers energy without requiring matter moving from one place to another.
2. Describe the four properties of waves: wavelength, frequency, velocity, and amplitude.
3. Define the equation that relates wavelength, frequency, and velocity.
4. Distinguish between transverse and longitudinal waves.
5. Describe the importance of electromagnetic radiation to the activities of living organisms.
6. Describe how waves can interfere with one another, both constructively and destructively.
7. Discuss the relationship of the velocity of electromagnetic radiation to the speed of light.
8. Describe the Doppler effect.
9. Discuss how electromagnetic radiation is transmitted, absorbed and scattered.
10. Discuss the extent of the electromagnetic spectrum, and discuss the importance of different types of waves to human technological developments.

Key Chapter Concepts

- A wave is a traveling disturbance that carries energy from place to place without requiring matter to travel across the intervening distance.
- Waves display certain properties:
  - Wavelength is the distance between crests or the highest points of adjacent waves.
  - Frequency is the number of wave crests that go by a given point every second. One wave crest per second (one cycle) equals one herz (1 Hz).
  - Velocity is the speed and direction of the wave crest itself.
  - Amplitude is the height of the wave crest above the undisturbed position.
- There are two kinds of waves:
  - Transverse waves, such as waves on a pond, move perpendicular to the motion of the medium through which they pass.
  - Longitudinal waves, such as sound waves, in which the medium moves in the same direction as the wave.
- Many animals use sound waves to communicate.
- Interference is produced when sound waves from different sources arrive at the same point.
  - Constructive interference results when both waves displace the medium in the same direction simultaneously.
  - Destructive interference results when the waves displace the medium in opposite directions simultaneously.
- An electromagnetic wave is a field of electric and magnetic energy that is self-perpetuating as it moves through space.
- All electromagnetic waves move at the speed of light, 300,000 km/s (= 186,000 miles/s).
- As the wavelength of the electromagnetic radiation changes, the characteristics of the radiation change.
- The Doppler effect describes the way the frequency of a wave appears to change if there is relative motion between the wave source and the observer.
  - If the source is moving toward the observer, the frequency appears to increase, as indicated by the rising pitch of an oncoming siren.
  - If the source is moving away from the observer, the frequency appears to decrease, as indicated by the lower pitch of the siren once it has passed the observer.
  - Light waves also display the Doppler effect.
- When electromagnetic wave hits matter, one of three things can happen:
  - It can be transmitted through the matter.
  - It can be absorbed by the matter.
  - It can be scattered (reflected) by the matter.
- The electromagnetic spectrum is very diverse, and ranges from waves with wavelengths longer than the radius of the earth to those with wavelengths shorter than the size of the nucleus of the atom.
  - Radio waves are the longest waves. Radio antennas accelerate electrons producing outgoing waves that are picked up by the antenna in your radio. The electronics in the radio transfer the varying signals to the speaker, which produces sounds. AM radio modulates the amplitude of the wave. FM radio modulates the frequency of the wave.
  - Microwaves are used to transmit radio and television signals from earth to satellites and back again. Microwave ovens focus microwaves into a food item, where they transmit their energy to water molecules in the food, thereby cooking it.

- Infrared radiation is given off in great quantities by the sun and felt as heat by the skin.
- The eye is sensitive to the wavelengths of visible light.
- Ultraviolet radiation is produced in great quantities by the sun. In small quantities, it causes tanning in the skin; in larger quantities, it causes extensive damage to DNA in the cells, often leading to cancer.
- X-rays have very short wavelengths, and can easily penetrate solid matter, such as most structures in the body.
- Gamma rays have the shortest wavelengths, and highest energies, of all electromagnetic radiation. Gamma rays often kill the cells they pass through, and if carefully regulated may be used to destroy tumors.

Key Individuals in Science

- James Clerk Maxwell (1831-1879) predicted the existence of electromagnetic radiation. He also discovered that light and other kinds of radiation are types of waves that are generated whenever electrical charges are accelerated.
- Albert Michelson (1852-1931) demonstrated that the hypothesized ether that occupied all of space did not exist.
- Edward W. Morley (1838-1923) demonstrated that the hypothesized ether that occupied all of space did not exist.
- Christian Johann Doppler (1803-1853) discovered that the frequency of an electromagnetic wave would be raised or lowered, depending on whether the source of the wave was moving toward or away from an observer, respectively.
- Heinrich Rudolph Herz (1857-1894) discovered radio waves, and demonstrated that the electromagnetic spectrum extended beyond the visible spectrum.

Key Formulas and Equations

- Wave velocity (m/s) = wavelength (m) x frequency (Hz)
- Wavelength (in m) = velocity (in m/s)
  $$\text{frequency (in Hz)}$$

- 1 hertz (Hz) = 1 cycle/second

- In a vacuum: wavelength (in m) x frequency (in Hz) = c

- $c = 300{,}000$ km/s $= 3 \times 10^8$ m/s $= 186{,}000$ miles/s

Key Concept: Waves Possess Four Properties

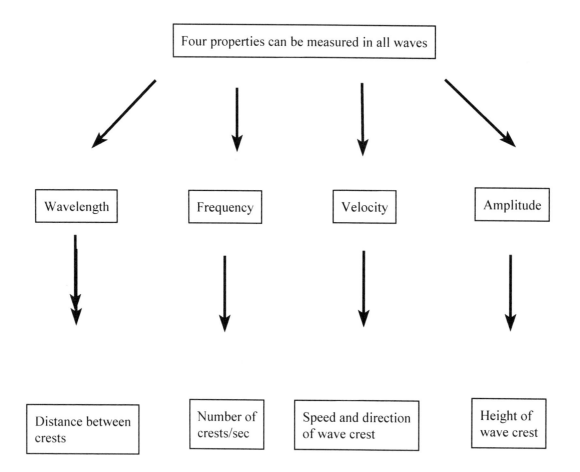

Key Concept: Animals Use Sound Waves

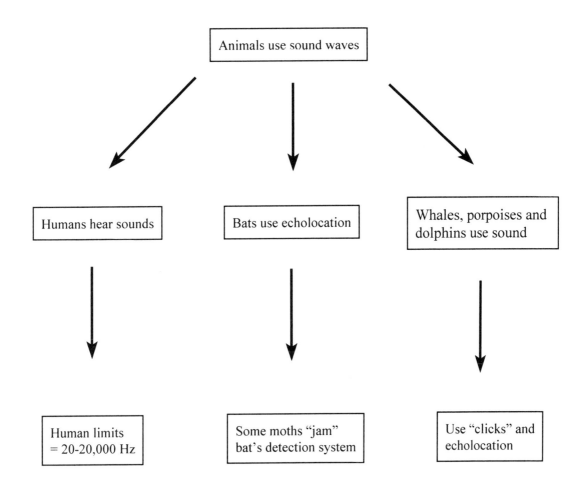

Key Concept: The Electromagnetic Spectrum

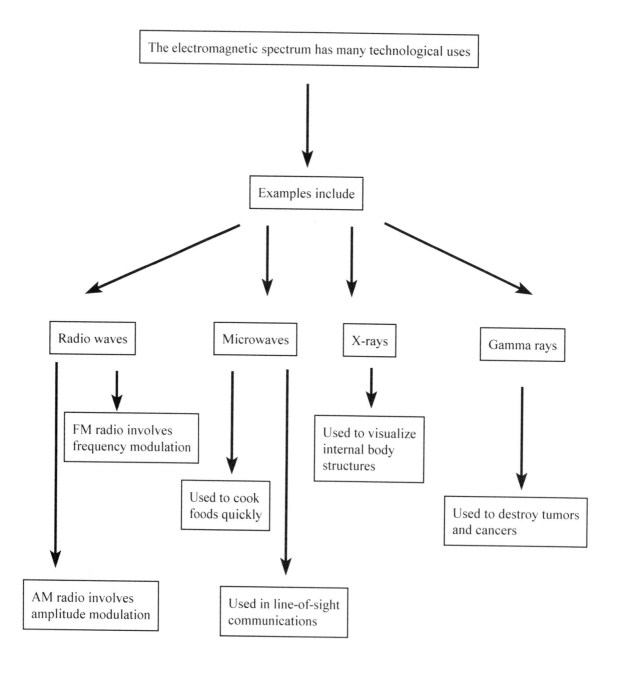

Questions for Review

Multiple-Choice Questions:

1. The distance between crests in a wave is its
    a. wavelength
    b. frequency
    c. velocity
    d. hertz rating

2. The height of a wave crest above the undisturbed position is its
    a. wavelength
    b. frequency
    c. velocity
    d. amplitude

3. In a transverse wave,
    a. the motion of the medium is parallel to the direction of the wave
    b. the motion of the medium is perpendicular to the direction of the wave
    c. the motion of the medium surrounds the direction of the wave
    d. the motion of the medium has no relationship to the direction to the wave

4. In a longitudinal wave,
    a. the motion of the medium is parallel to the direction of the wave
    b. the motion of the medium is perpendicular to the direction of the wave
    c. the motion of the medium surrounds the direction of the wave
    d. the motion of the medium has no relationship to the direction to the wave

5. Bats use high-pitched sound waves undetectable to human hearing to detect objects in their environment. This means the wavelength of the sounds they use is
    a. between 20 and 200 Hz
    b. between 200 and 1,000 Hz
    c. between 2,000 and 20,000 Hz
    d. above 20,000 Hz

6. The velocity of all electromagnetic radiation is
    a. dependent on the type of radiation
    b. the speed of light
    c. the speed of sound
    d. faster for the longest wavelengths

7. The increase in frequency of an approaching source of electromagnetic radiation is termed the
    a. absorption spectrum
    b. the transmission factor
    c. the scattering effect
    d. the Doppler effect

8. If you are standing in the path of a source of light that is moving toward you, the light you see will look _____ than it would ordinarily.
    a. redder
    b. bluer
    c. brighter
    d. darker

9. When light passes through a transparent material, its direction is usually changed slightly. This is called
    a. reflection
    b. scattering
    c. refraction
    d. absorption

10. Which of the following has the shortest wavelength?
    a. fm radio waves
    b. am radio waves
    c. microwaves
    d. gamma rays

Fill-In Questions:

11. _____ rays are used in medicine to destroy tumors and cancers.
12. A sunburn is the result of exposure to _____ radiation in sunlight.
13. Heat can usually be detected as _____ radiation.
14. A frequency of one cycle per second in a wave is called one _____.
15. _____ is the number of wave crests that go by a given point every second.
16. Constructive _____ causes an additive effect when two waves reach the same point at the same time.
17. The speed of light is _____ km/s.
18. In order to determine the frequency of an electromagnetic radiation, you would divide the velocity by the _____.
19. _____ occurs when waves from two different sources come together at a single point.
20. The distance between crests in a wave is its _____.

Problems:

1. If we know that a seismic (earthquake) wave has a velocity of 5,000 m/s and its frequency is 25 Hz; what is its wavelength in meters?

2. If an earthquake occurs in Death Valley, California, with a wavelength of 2,000 m and a frequency of 5 Hz. How long will it take the residents in downtown Los Angeles, 300 km away, to feel the temblor?

3. What is the frequency of electromagnetic radiation from outer space that reaches the earth with a wavelength of 1×10^5 m, and a velocity of 1×10^5 m/sec?

Crossword Quiz: Electromagnetic Radiation

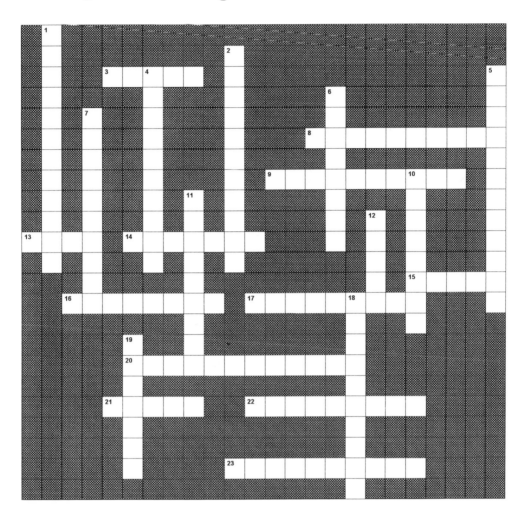

ACROSS

3. The unit of measurement for the frequency of waves
8. This happens when electromagnetic waves slow down and change direction
9. A process by which electromagnetic waves may be absorbed and rapidly re-emitted
13. A traveling disturbance that carries energy from place to place
14. The change in frequency or wavelength of a wave detected by an observer due to the movement of the source of the wave
15. These waves may be commercially modified either by ampitude or frequency
16. The speed of travel of electromagnetic waves
17. Electromagnetic waves with wavelengths ranging from approximately 1 meter to 1 millimeter, used in line-of-sight communications

20. The interaction of waves from two different sources arriving at a single point
21. The speed of _____ is 300,000 km/s
22. The number of wave crests that go by a given point every second
23. The conversion of electromagnetic energy into heat energy

DOWN

1. The kind of wave in which the motion of the medium is in the same direction as the wave movement
2. A type of radiation that causes skin cancer
4. The scattering of light waves from the surface of an object
5. The phenomenon of light energy passing through matter unaffected
6. The range of wavelengths of electromagnetic radiations in the universe
7. A kind of wave in which the motion of the wave is perpendicular to the motion of the medium
10. Wavelengths of electromagnetic radiation felt as heat radiation
11. The height of a wave crest above the undisturbed level of the medium
12. The highest energy wave of the electromagnetic spectrum
18. The distance between crests of adjacent waves
19. The portion of the electromagnetic spectrum that is detectable by the human eye

Answers to Review Questions

Multiple-Choice Questions

1. a; 2. d; 3. b; 4.a; 5.d; 6. b; 7. d; 8. b; 9. c; 10. d

Fill-In Questions:

11. gamma; 12. ultraviolet; 13. infrared; 14. Hertz (Hz); 15. frequency; 16. interference; 17. 300,000; 18. wavelength; 19. interference; 20. wavelength

Chapter 7

Albert Einstein and the Theory of Relativity

Chapter Review

Albert Einstein, a brilliant physicist and mathematician, noted that different observers of the same event from different perspectives have a different frame of reference. In his special principle of relativity, he stated that even though different observations appear to rely on the frame of reference, all observers are measuring the same universal laws of nature. Special relativity deals with observers who are not accelerating with respect to one another, while general relativity deals with observers in any frame of reference whatsoever. One of the tenets of the principle of relativity is that the speed of light is the same everywhere in the universe, and in all reference frames. A consequence of the principle of relativity is that time as measured on a clock is relative to one's frame of reference. Time measured on a moving clock is expanded relative to time measured on a stationary clock, that is the moving clock appears to tick more slowly than the stationary one. As the speed of light is approached, time slows down and approaches zero. During everyday experiences, we are unaware of the effects of relativity. Only when the velocity of the moving clock approaches the speed of light does the dilation of time become readily apparent. If space travel ever approaches the speed of light, the astronauts would age much slower than the people left back on earth. A second consequence of relativity is that an object traveling at near the speed of light shortens along the direction of travel. As the speed of light is approached, distances shrink and approach zero. A third consequence of relativity is that moving objects become more massive. As the velocity of a moving object approaches the speed of light, its mass approaches infinity. The principle of relativity also led to the development of the equation $E = mc^2$. The general principle of relativity states that acceleration and gravity are related to one another. In this universe, heavy masses warp the fabric of space-time and affect the motion of other objects. Three observations of general relativity involve the bending of light rays as they pass near the sun, changes in the orbit of the planet Mercury, and the shift in frequency and wavelength of light passing through a gravitational field.

Learning Objectives

After studying this chapter, you should be able to:
(Other objectives may also be assigned by your instructor)

1. Describe how the same event can appear different to an observer who is stationary and one who is moving.
2. State Einstein's principle of relativity, and distinguish between special relativity and general relativity.

3. Indicate the importance of the speed of light to the principle of relativity.
4. Explain how time stretches out when it is measured by a moving observer with a moving clock.
5. Describe the Lorentz factor and indicate its importance to calculating the changes predicted by the principle of relativity.
6. Explain why distances shrink as one approaches the speed of light.
7. Explain why the mass of an object increases as that object's velocity approaches the speed of light.
8. Discuss Einstein's famous equation $E = mc^2$.
9. Describe how the general principle of relativity links acceleration and gravity, and cite examples of convincing evidence.

Key Chapter Concepts

*   The frame of reference from which observations of the universe are made have a profound effect on the observations.
    *   A person may be stationary and observe events on a stationary platform near the person.
    *   A person may be moving with a platform and make observations on stationary events as the moving platform passes by.
    *   A moving person may observe events occurring on the moving platform that is also carrying the person.
    *   Einstein's principle of relativity states that, even though an event may appear to be different, depending upon whether the observer is moving relative to the event, every observer must experience the same natural laws.
    *   Special relativity deals with all frames of reference in uniform motion relative to one another.
    *   General relativity deals with any reference frame whether or not it is accelerating relative to the observer.
    *   The speed of light, c, is the same in all reference frames.
*   Time varies in its passage depending on whether it is being measured on a stationary clock or a moving clock.
    *   Time intervals remain the same in a stationary clock.
    *   Time intervals are dilated (stretched out) in a moving clock. The faster the clock moves, the greater the dilation. As the clock approaches the speed of light, time intervals become progressively longer, and at the speed of light, time intervals are infinitely long.
    *   If space travel could ever approach the speed of light, time would slow down on the spacecraft, relative to how fast time was passing on the earth, and the astronauts aboard would age at less than one-tenth the rate of their relatives on earth.
*   Distances shrink and the length of a moving object contracts when its velocity approaches the speed of light. A basketball moving at a speed near the speed of light would take on the shape of a pancake.
*   The mass of an object is relative to its speed.
    *   The faster an object travels, the more massive it is. As the speed of an object approaches the speed of light, its mass approaches infinity.
    *   The only objects that can travel at the speed of light (photons) are those that have zero rest mass.
*   The rest energy of an object equals its rest mass times the speed of light squared. During nuclear fission and nuclear fusion reactions, large quantities of energy are released when small quantities

of mass are converted directly into energy.

- The principle of general relativity states that acceleration and gravity are equivalent forces.
  - This principle describes the space-time continuum as distorted by the mass of celestial bodies into a curved entity.
  - Evidence for the principle of general relativity has been gathered in three observations.
    - Light coming from a distant star is bent toward the sun by the sun's gravitational field.
    - Einstein's principles accurately predicted small, but detectable changes in the orbit of the planet Mercury.
    - The principle of relativity explains the shift in frequency and wavelength observed when light passes through a gravitational field.
- Einstein's theories do not contradict Newton's theories studied in Chapter 2. Newton's laws deal with objects traveling at speeds much slower than the speed of light, and under these circumstances, both Newton and Einstein correspond.

Key Individuals in Science

- Albert Einstein published the theory of relativity in 1905.

Key Formulas and Equations

- time = $\dfrac{\text{distance}}{\text{speed}}$, $t = \dfrac{d}{c}$, where c = the speed of light

- The Lorentz factor = the square root of $[1-(v/c)^2]$ or $\sqrt{1-(v/c)^2}$

- To calculate the difference in time between two clocks: a moving clock (t_{MG}) and a stationary clock (t_{GG}) observed by the same person:

$$t_{MG} = \frac{t_>}{\sqrt{1-(v/c)^2}}$$

- To calculate the length (L) contraction of an object approaching the speed of light:

$$L_{MG} = L_> x\sqrt{1-(v/c)^2}$$

- To calculate the increase in mass of an object approaching the speed of light:

$$m_{MG} = \frac{m_>}{\sqrt{1-(v/c)^2}}$$

Key Concept: Concepts of Special Relativity

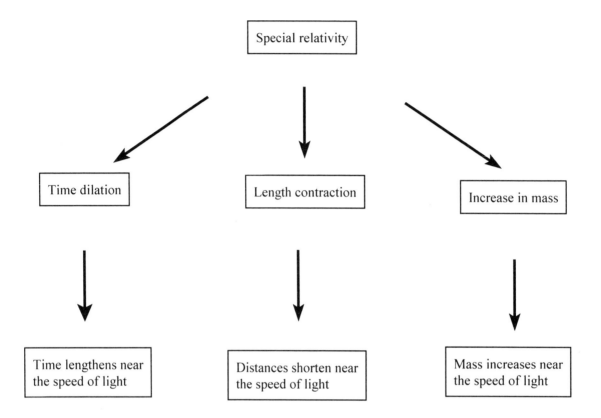

Questions for Review

Multiple-Choice Questions

1. Consider a man throwing a ball forward from the front of a moving train. If the train is moving at 50 km/hour and the man throws the ball at 5 km per hour, what is the speed apparent to the thrower?
    a. 55 km/hour
    b. 100 km/hour
    c. 5 km/hour
    d. 105 km/hour

2. If another person is standing and watching alongside the road in front of the train, what is the apparent velocity of the ball coming toward this observer?
    a. 55 km/hour
    b. 100 km/hour
    c. 5 km/hour
    d. 105 km/hour

3. If a person is watching this from behind the train, what velocity would he measure?
    a. 55 km/hour
    b. 100 km/hour
    c. 5 km/hour
    d. 105 km/hour

4. The special theory of relativity deals with
    a. all frames of reference in the universe
    b. all frames of reference in uniform motion relative to one another
    c. any reference frame whether or not it is accelerated relative to another
    d. none of the above

5. Einstein's relativity theories require that one factor is the same in the entire universe at all times. It is:
    a. the size of the nucleus
    b. the speed of oncoming object
    c. the reference point of the observer
    d. the speed of light

6. The general theory of relativity deals with
    a. linear frames of reference in the universe
    b. all frames of reference in uniform motion relative to one another
    c. any reference frame whether or not it is accelerated relative to another
    d. none of the above

7. If a moving clock is registering time when it approaches the speed of light, the time it registers will be
   a. much faster than a clock in a stationary position
   b. much slower than a clock in a stationary position
   c. the same as a clock in a stationary position
   d. unable to measure time accurately

8. If an object approaches the speed of light, its length will
   a. shorten
   b. remain the same
   c. lengthen
   d. be transferred to its width

9. The mass of an object that approaches the speed of light will
   a. increase
   b. decrease
   c. remain the same
   d. none of the above

10. When matter is transformed into energy in a nuclear fission reaction, the constant factor in this reaction is
    a. the number of protons present
    b. the number of neutrons present
    c. the number electrons present
    d. the speed of light squared

Fill-In Questions

11. Albert Einstein is known because he challenged the laws of physics first promoted by _____.
12. The position of an observer measuring an outside event is referred to as his/her _____.
13. The principle of _____ explains the bending of light when it passes through a gravitational field.
14. The single constant that can be measured throughout the universe is the _____.
15. When an object approaches the speed of light, its length is _____.
16. In relativistic equations, time equals distance divided by _____.
17. The expression ·· · ··· is known as the _____.
18. The mass of an object increases as it approaches the _____.
19. An object that is moving near the speed of light appears _____ than an object traveling at a slower speed.
20. According to Einstein, rest mass x the speed of light squared = _____.

There is no Crossword Quiz in this chapter.

Answers to Review Questions

Multiple-Choice Questions

1. c; 2. a; 3. a; 4. b; 5. d; 6. c; 7. b; 8. a; 9. a; 10. d

Fill-In Questions

11. Newton; 12. frame of reference; 13. general relativity; 14. speed of light; 15. shortened; 16. speed; 17. Lorentz factor; 18. speed of light; 19. shorter; 20. rest energy

Chapter 8

The Atom

Chapter Review

Atoms are the building blocks of the universe. All of the elements, both naturally occurring, and artificially manufactured by scientists are composed of atoms. An atom consists of a massive central nucleus that contains positively charged protons and electrically neutral neutrons. Negatively charged electrons orbit the nucleus, occupying various energy levels. An electron can shift from lower energy orbits to higher energy orbits by absorbing energy in the form of heat or light. When an electron shifts to a lower energy level, it emits energy in the form of electromagnetic radiation. Such energy shifts are called quantum leaps. The distribution of electrons around the atom, particularly those in the outermost orbits, is responsible for the chemical reactions of every type of atom. Scientists have discovered 92 different types of atoms that occur naturally in the universe. In addition, scientists have been able to make several new types of atoms. Each different atom is called an element. A group of two or more atoms chemically bound together is called a molecule.

Learning Objectives

After studying this chapter, you should be able to:
(Other objectives may also be assigned by your instructor)

1. Define an atom and describe the different components that comprise an atom.
2. Discuss the relationship between atoms and elements.
3. Distinguish between the two hypothetical models for the atom proposed by Ernest Rutherford and Niels Bohr.
4. Define a photon of light and describe its importance in quantum leaps.
5. Discuss the discovery that, when excited, the atoms of different elements emitted different wavelengths of light.
6. Describe how a laser beam is produced.
7. Describe how the periodic table of the elements organizes the known types of elements into groups with similar chemical properties.
8. Define the concept of a mole, and indicate how it is important when describing quantities of atoms.

Key Chapter Concepts

*   A number of scientific observations provide evidence for the existence of atoms:
    *   The behavior of gases under pressure is best explained by the collisions of separate atoms.
    *   The fact that elements always combine in definite proportions based on their atomic weights

is evidence for the existence of atoms.
  - The emission of radiation from individual atoms is strong evidence for the existence of atoms.
  - Albert Einstein's mathematical analysis of the erratic motion (Brownian motion) of dust particles suspended in water argued strongly for the existence of atoms.
  - X-ray crystallography provided visual proof of the arrangement of atoms in crystals.
  - Scanning tunneling microscopes can produce images of individual atoms.
- An element is a substance composed of atoms of only one kind.
  - An atom is the smallest unit of an element having the chemical characteristics of that element. An element cannot be broken down into units smaller than its atoms
  - There are 92 naturally occurring elements in the universe, and more elements have been made in the laboratory by scientists.
  - Elements can be arranged into related groups based on their chemical characteristics. This grouping is called the Periodic Table of the Elements.
  - A mole of a substance is the amount of the substance that consists of Avogadro's number (6.02×10^{23}) of atoms or molecules, and the weight of one mole of substance is equal to the atomic or molecular weight of that substance
- All atoms have a consistent structure, including a central nucleus surrounded by a cloud of electrons.
  - The nucleus of an atom is a small, massive, centrally located structure containing positively charged protons and electrically neutral neutrons.
  - Electrons may exist in one of several orbits, or energy levels, surrounding the nucleus of the atom.
  - Whenever an electron changes orbits, it either must absorb energy to move to a higher energy level (orbit) closer to the nucleus , or give off energy to move to a lower energy level farther away from the nucleus.
  - The energy absorbed or released by electrons as they move from one energy level to another is measured in photon units.
  - A photon is a unit of electromagnetic radiation that can be absorbed or emitted by an electron.
  - The movement of electrons from one orbit, or energy level, to another is termed a quantum leap or quantum jump.
- The science of spectroscopy is used to study the quantity and quality of photons released by electrons from elements, when those electrons are excited.
  - Electrons that move from one energy level (a quantum leap) to another give off photons.
  - The electrons of every element will emit a different array of photons when they undergo quantum leaps.
  - A spectroscope passes light emitted from excited electrons through a prism, and analyzes the specific photons emanating from the sample.
  - The spectroscope allows scientists to identify exactly which element, whether in outer space or on earth, has emitted a particular spectrographic signature.
- Lasers make use of the ability of electrons to absorb and emit photons of light.
  - Atoms properly stimulated will emit photons in a special way.
  - Mirrors are used to amplify the production of coherent beams of light.
  - Some light is allowed to escape the system as the familiar laser beam.
  - Laser beams have many uses in science and technology, especially in making extremely precise and detailed measurements of the properties and structures of atoms.

Key Individuals in Science

- In the sixth century, B.C., Democritus, a Greek philosopher, introduced the concept of the atom, an indivisible piece of matter.
- John Dalton (1766-1844) provided the modern concept of atoms being the building blocks of elements.
- Joseph John Thomson (1856-1940) first identified the electron.
- Ernest Rutherford (1871-1937) performed experiments that showed that atoms consisted of a small, dense nucleus surrounded by light electrons orbiting the nucleus.
- Niels Bohr (1885-1962) introduced the concept of specific energy levels that were occupied by electrons.
- Joseph Norman Lockyer (1836-1920) discovered the element helium by studying spectrographic analyses of light from the sun.
- Dimitri Mendeleev (1834-1907) discovered the relationship of the atomic weight of an element to its chemical properties. This discovery allowed him to arrange the elements in a Periodic Table of the Elements.

Key Concept: The Structure of the Atom

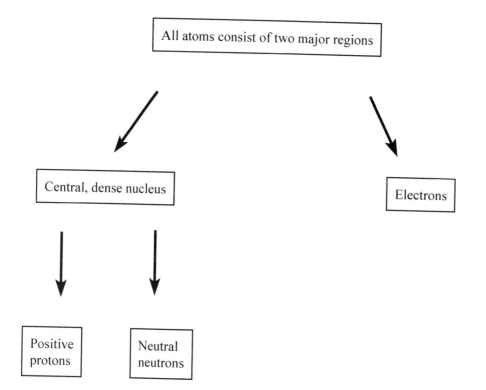

Key Concept: Electrons Make Quantum Leaps

| Electrons exist in different energy (quantum) levels |

↓ ↓

| If an electron absorbs energy | | If an electron emits energy |

↓ ↓

| Electron jumps to a higher energy level | | Electron jumps to a lower energy level |

Questions for Review

Multiple-Choice Questions:

1. According to Democritus, all material was formed from
    a. molecules
    b. atoms
    c. fire and earth
    d. air and water

2. John Dalton proposed the hypothesis of
    a. laser beams
    b. quantum leaps
    c. the periodic table of the elements
    d. atomism

3. Two or more atoms stuck together are a(n)
    a. element
    b. molecule
    c. double atom
    d. nucleus

4. The dense, central portion of an atom is its
    a. quantum group
    b. electron
    c. photons
    d. nucleus

5. Helium, gold, carbon, aluminum and copper are examples of
    a. elements
    b. atoms
    c. molecules
    d. quantum groups

6. The arrangement of elements into a periodic table was originally proposed by
    a. Mendeleev
    b. Democritus
    c. Dalton
    d. Bohr

7. The science of spectroscopy examines the nature of _____ emitted from excited atoms.
    a. electrons
    b. protons
    c. photons
    d. neutrons

8. The positively charged particle found in the nucleus of an atom is
    a. proton
    b. neutron
    c. electron
    d. photon

9. The total collection of photons emitted by a given atom is called its
    a. energy signature
    b. spectrum
    c. periodic position
    d. electromagnetic image

10. The negatively charged particle that is a part of an atom is
    a. electron
    b. proton
    c. neutron
    d. photon

Fill-In Questions:

11. The Greek philosopher who coined the term "atom" was _____.
12. If a chemist uses an electrical current to break down water, two gases are produced: hydrogen and
    _____.
13. A material that cannot be broken down into other such materials is an _____.
14. Three atoms joined together are called a _____.
15. The periodic table of the elements was proposed by _____.
16. An _____ is an electrically charged atom.
17. The smallest particle that can retain its chemical identity is the _____.
18. The _____ is a positively charged nuclear particle.
19. A _____ is an electrically neutral nuclear particle.
20. The science that studies how atoms absorb and emit electromagnetic radiation is _____.

Crossword Quiz: The Atom

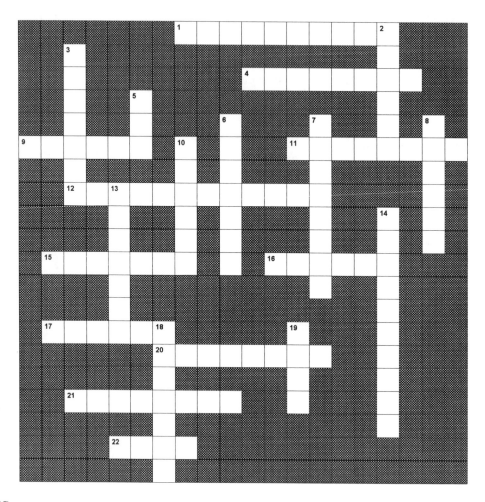

ACROSS

1. He discovered the existence of the atomic nucleus
4. Two or more atoms joined together
9. The unit of electromagnetic energy that is emitted from an atom when an electron changes energy levels
11 This type of erratic motion is seen in dust particles suspended in water
12 A process whereby electricity is used to break molecules down into atoms
15. He proved that electrons were electrically charged particles
16. An element found in the form of a light gas
17. Positively charged particles found in the nucleus of the atom
20. Negatively charged particles that orbit the nucleus of an atom
21. The array of electromagnetic energy that is absorbed or emitted by an element
22. The indivisible unit of an element

DOWN

2. He discovered the existence of elements
3. A metal plate in a battery that is positively charged
5. An electrically charged atom
6. The dense, central core of the atom
7. X-rays can be used to determine the arrangement of atoms in _____
8. This is the black element produced when wood is burned
10. He experimentally confirmed Einstein's statistical predictions about the movement of atoms in Brownian motion
13. A material that cannot be broken down into smaller such materials
14. Greek philosopher who first used the term "atom"
18. Electrically neutral particles found in the nucleus of the atom
19. He described the positions of electrons in different energy states

Answers to Review Questions

Multiple-Choice Questions:

1. b; 2. d; 3. b; 4. d; 5. a; 6. a; 7. b; 8. a; 9. b; 10. a

Fill-in Questions:

11. Democritus; 12. oxygen; 13. element; 14. molecules; 15. Mendeleev; 16. ion; 17. atom; 18. proton; 19. neutron; 20. spectroscopy.

Chapter 9

Quantum Mechanics

Chapter Review

The Newtonian principles governing the movement of objects that you studied in Chapter 2 are useful in understanding how large objects accelerate and are affected by gravity. However, objects as small as atoms, and subatomic particles, such as electrons, do not follow the same laws as large objects. Different rules, called quantum mechanics, allow us to describe and predict events at the subatomic level. The Heisenberg uncertainty principle takes into account the fact that merely measuring the position or movement of subatomic particles can significantly change their behavior. This principle teaches us that we can never accurately predict the exact position or velocity of these subatomic particles. Instead, we can only estimate the probability of the position or velocity of the particle at any specific point in time. Further study of the movements of these particles shows that they are capable of behaving as both objects (particles) and waves simultaneously.

Learning Objectives

After studying this chapter, you should be able to:
(Other objectives may also be assigned by your instructor)

1. Define the concept of how matter, energy, and electrical charge are quantized in the atom.
2. Discuss how the study of quantum mechanics is used to describe the uncertainty of the position or velocity of quantized particles.
3. Discuss the techniques that scientists use to study subatomic particles, since they are too small to be seen.
4. Define the Heisenberg, or uncertainty, principle, and tell how it deals with uncertainty in quantum mechanics.
5. Describe how probability is used to describe the position or velocity of subatomic particles.
6. Discuss the concept of the wave-particle duality exhibited by subatomic particles.
7. Describe the double-slit test and indicate how it supports the wave-particle duality.
8. Describe how photons can be used to generate electrical currents, and the practical uses of these currents.

Key Chapter Concepts

*   Quantum mechanics is the study of the motion of tiny pieces of matter, such as electrons, protons and neutrons.
    *   Electrical charge, and mass of subatomic particles comes in exact, small bundles, called

quanta. Physicists say the atom's matter and electrical charge is quantized.
- Quantum mechanics is not used to explain the location or movement of large objects, like apples and automobiles.
- Measurements made on objects that contain less energy than the measuring device do alter the object.
- Measurements made on quantized objects significantly changes the object being measured.
- The Heisenberg uncertainty principle tells us that the error of uncertainty in the measurement of an object's position, times the error or uncertainty in that object's velocity, must be greater than a constant (Planck''s constant) divided by the object's mass.
    - The larger an object, the less subject it is to the uncertainties of the Heisenberg principle.
    - The smaller an object, the more subject it is to the Heisenberg principle.
    - One of the implications of the Heisenberg principle is that if an object is very small, we can only designate the probability that it will be in any area when we expect it to be.
- Most small particles may behave as either an electromagnetic ware or a particle.
    - In practice, photons can behave either as waves or particles, depending on the type of experiment you perform.
    - Photons may cause the release of electrons from some surfaces with which they collide.
    - The electrical energy release may be used to create images, CAT scans, of the material penetrated by the electrons.
    - The study of subatomic particles demands scientists to be able to deal with phenomena that are not duplicated in large objects.

Key Individuals in Science

- Werner Heisenberg (1901-1976) formulated the Heisenberg uncertainty principle, which states that at the quantum scale, any measurement significantly alters the object being measured.

Key Formulas and Equations

- The Heisenberg uncertainty principle states that the error or uncertainty in the measurement of an objects position, times the error of uncertainty in that object's velocity, must be greater than constant (Planck's constant) divided by the object's mass.
    - (uncertainty in position) x (uncertainty in velocity) $> \dfrac{h}{\text{mass}}$

    - $\Delta x \times \Delta v > \dfrac{h}{m}$

Key Concept: The Heisenberg Uncertainty Principle

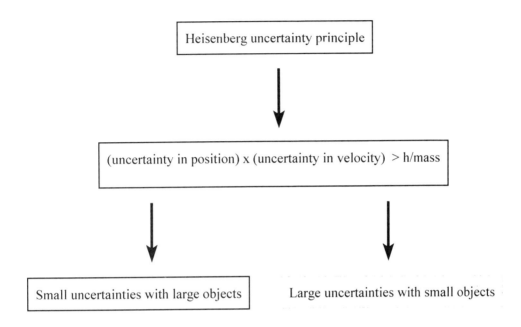

Questions for Review

Multiple-Choice Questions

1. A quantum leap occurs when a(an) _____ moves between energy levels in an atom.
    a.   proton
    b.   neutron
    c.   electron
    d.   photon

2. Every measurement in the physical world incorporates all of the following <u>except</u>:
    a.   photons
    b.   a sample
    c.   a source of energy
    d.   a detector

3. The principle of uncertainty was originally proposed by
    a.   Dalton
    b.   Heisenberg
    c.   Mendeleev
    d.   Democritus

4. When we use the uncertainty principle to calculate the relationship between the position and velocity of a subatomic particle, as we become precise about the position of a particle,
    a.   we become more certain about its velocity than its position
    b.   we become equally certain about its velocity and position
    c.   we become less certain about its velocity than its position
    d.   its velocity has no effect on its position

5. Because of the uncertainty principle, scientists are forced to rely on _____ to describe the position of a small object whose velocity is known.
    a.   probability
    b.   Newton's laws of motion
    c.   the wave-particle duality principle
    d.   hypothetical estimates

6. The double-slit experiment was important in documenting
    a.   the small uncertainty of large objects
    b.   the large uncertainty of small objects
    c.   the wave-particle duality principle
    d.   none of the above

7. The production of an electrical charge when light strikes the surface of an object is called the
   a. wave-duality principle
   b. photoelectric effect
   c. quantum leap theory
   d. Heisenberg principle

Fill-In Questions

8. The study of the motion of small objects that come in little bundles is called _____.
9. _____, a German physicist, propounded the uncertainty principle.
10. In the uncertainty equation, the term "h" stands for _____.
11. The chance that an object will be in a certain place at a certain time is called _____.
12. Quantum mechanics is sometimes called _____ mechanics.
13. The _____ test was used to validate the wave-particle duality hypothesis.
14. When an electron moves from one energy level to another, light is emitted in discrete bundles of energy called _____.
15. During the photoelectric effect, when _____ strikes some materials, their energy can force electrons to emerge from the opposite surface of the material.
16. Energy in the form of _____ is used to produced a CAT scan.
17. In addition to the movement of a particle, the motion of an electron can also be thought of as that of a _____.
18. Amplification of the intensities of adjacent wavelengths by electrons moving through a two-slit barrier is an example of _____ interference.

There is no Crossword Quiz for Chapter 8.

Answers to Review Questions

Multiple-Choice Questions

1. c; 2. a; 3. b; 4. c; 5. a; 6. c; 7. b;

Fill-In Questions

8. quantum mechanics; 9. Heisenberg; 10. Planck's constant; 11. probability; 12. wave; 13. double-slit; 14. photons; 15. photons; 16. X-rays; 17. wave; 18. constructive

Chapter 10

Atoms in Combination: The Chemical Bond

Chapter Review

Chemical bonds are the means used by atoms to connect together. The incentive to form chemical bonds lies in the fact that the association of atoms thus formed lowers the potential energy of the chemical system, and provides more stability for the group of atoms formed. Ionic bonds involve the transfer of electrons from one atom to another. In this type of bond, the positively charged ion that has lost one or more electrons is held closely to the negatively charged ion that has accepted those electrons by electrostatic forces. Metallic bonds are ionic bonds in which the electrons are shared among the participant atoms in the arrangement. Covalent bones involve two or more atoms sharing electrons, resulting in the formation of a molecule. Both hydrogen bonding and van der Waals forces involve the electrical attraction between atoms or molecules that are technically electrically neutral, but whose electrical charge is so distorted that certain parts of the molecule are slightly more positive or negative, and thereby capable of forming weak electrostatic attractions between opposites. Matter is found is several different states. Solids have a fixed shape and volume, and include crystals, glasses, and plastics. Liquids have a fixed volume but no fixed shape. The molecules in gases have enough kinetic energy to fill the space within which they are enclosed. Many materials, such as water, have the ability to change state, undergoing freezing, melting, and boiling. Chemical reactions, the interactions between atoms and molecules, may involve making or breaking chemical bonds, or simply rearranging the positions of atoms in a molecule. Two of the more common chemical reactions are oxidation, in which electrons are removed from an atom, and reduction in which electrons are added to an atom. Polymerization reactions involve the linking together of long chains of small molecules to form extremely large molecules, such as polyesters, nylon, vinyl, silk, and plastics. Depolymerization is the breaking down of long polymers into their constituent parts.

Learning Objectives

After studying this chapter, you should be able to:
(Other objectives may also be assigned by your instructor)

1. Define a chemical bond and distinguish among ionic, covalent, and hydrogen bonds.
2. Describe how metallic bonds differ from ordinary ionic bonds.
3. Discuss the differences between the transfer of electrons between atoms and the sharing of electrons between atoms.
4. Describe how a difference in polarity between two atoms can produce a hydrogen bond.
5. Describe van der Waals forces.
6. Distinguish among the different states of matter.

7. Define oxidation and reduction reactions in terms of the movement of electrons between the atoms involved.
8. Describe other chemical reactions, such as precipitation-solution reactions, acid-base reactions, and polymerization reactions.
9. Cite examples of how chemical reactions are essential to life.
10. Describe how to balance a simple chemical reaction.

Key Chapter Concepts

- The tendency for atoms to form chemical bonds is related to the number of electrons in their outer shells.
    - The electrons in an atom's outer shell are termed its valence electrons.
    - The most stable arrangement of electrons in an atom is to have a completely filled outer shell. This arrangement results in the state of lowest energy for the atom.
    - Atoms have a tendency to react in such a manner that they will achieve filled outer electron shells, and thus achieve their lowest energy state.
- A chemical bond allows two or more atoms to redistribute their electrons in such a way as to fill their outer electron shells, and thus arrive at the lowest energy state.
- There are three major kinds of chemical bonds: ionic, covalent, and hydrogen bonds.
- An ionic bond results when one or more electrons are transferred from one atom to another.
    - The transfer of electrons leaves the donor with unbalanced protons in the nucleus and a positive charge, while the recipient has unbalanced electrons and a negative charge.
    - The electrostatic charges among the atoms in such an arrangement tends to hold the charged atoms (ions) together in a crystalline configuration.
    - Metallic bonds are special cases of ionic bonds in which the electrons are shared by all of the atoms in the material.
- A covalent bond results when two or more atoms share electrons.
    - In the simplest covalent bonds, two atoms share a pair of electrons in their outermost shell in what is called a single bond.
    - Double and triple bonds may result from the sharing of two or three pairs of electrons, respectively.
    - Carbon, the element that forms the basis of life on earth, is capable of forming long chains, branching, and ring-shaped molecules, and three-dimensional frameworks in almost any imaginable shape, using a variety of covalent bonds.
- Hydrogen bonds are formed between the negatively polarized regions of certain molecules and slightly positively polarized hydrogen atoms in nearby molecules.
- Van der Waals forces are weak attractive forces between otherwise nonpolar or electrically neutral molecules.
- Matter occurs in several states: gases, plasmas, liquids, and solids.
    - A gas is any collection of atoms or molecules that expands to take the shape of and fill the volume available in its container.
    - A plasma is a state of matter in which positive nuclei move about in a sea of electrons.
    - A liquid is a collection of atoms or molecules that has no fixed shape but maintains a fixed volume.
    - A solid possesses a more or less fixed shape and volume.
        - Crystals are solids in which groups of atoms occur in a regularly repeating sequence in a predictable way.

- Glasses are solids with predictable local environments for most atoms, but no long-range order to the atomic structure.
    - Polymers are solids, such as plastics, animal hair, plant cellulose, cotton and spider webs, and are extremely long and large molecules that are formed from numerous smaller molecules, like links forming a chain.
  - Many materials may undergo changes in state.
    - Freezing and melting involve changes in state between liquids and solids.
    - Boiling and condensation are changes between liquids and gases.
    - Sublimation is the direct transformation of a solid into a gas.
- During chemical reactions, atoms or smaller molecules come together to form larger molecules, and larger molecules break up into smaller units.
  - During exothermic reactions, heat is released.
  - During endothermic reactions, heat is absorbed.
  - Oxidation includes any chemical reaction in which an atom loses electrons while combining with other elements. The atom that loses the electrons is said to be oxidized. Rusting and combustion are examples of oxidation reactions.
  - Reduction includes any chemical reaction in which an atom accepts electrons while combining with other elements. The atom that receives the electrons is said to be reduced.
  - During precipitation-solution reactions, a liquid, such as water, dissolves a solid, such as salt or sugar.
  - Acids are molecules that release positively charged hydrogen ions (H^+)when dissolved in water.
  - Bases are molecules that release negatively charged hydroxide ions (OH^-) when dissolved in water.
  - During polymerization reactions, numerous smaller molecules (monomers) are linked together in a repeated fashion to produce a large polymer. A common polymerization reaction is called condensation because a molecule of water is removed at each linkage. Depolymerization is the breakdown of a polymer into short segments. The clotting of blood is a example of polymerization in the human body.
  - Hydrocarbons, molecules composed of hydrogen and carbon, participate in numerous kinds of polymerizations.

Key Individuals in Science

- Wallace Carothers (1896-1937), led a team of chemists at the Du Pont chemical company who used polymerization to produce many commercially important products, including nylon and the synthetic rubber, neoprene.

Key Concept: Chemical Bonds are Attractive Forces Between Atoms

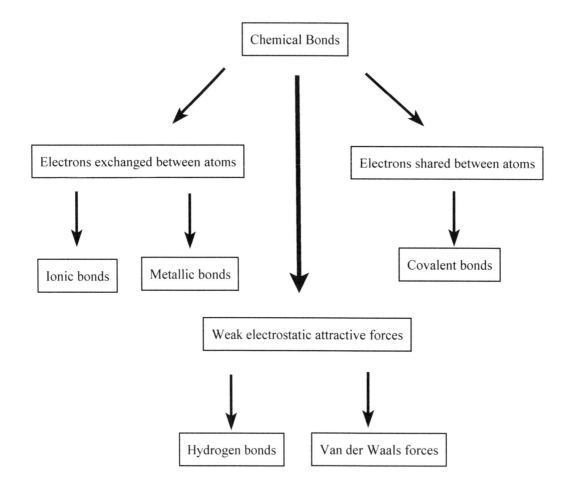

Key Concept: Matter is Found in Different Modes of Organization

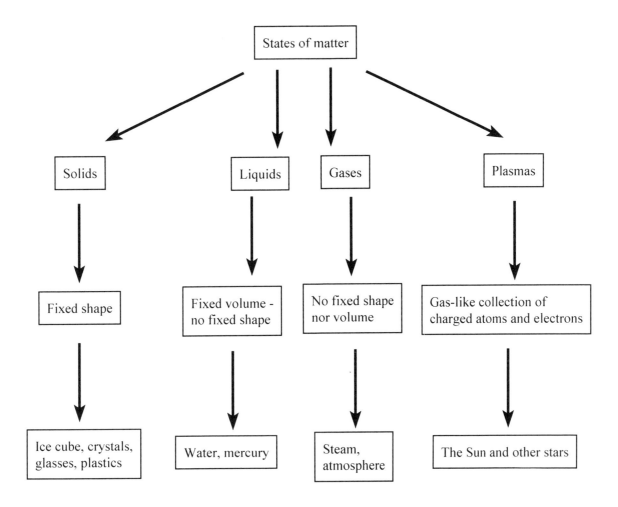

Key Concept: Atoms and Molecules Interact

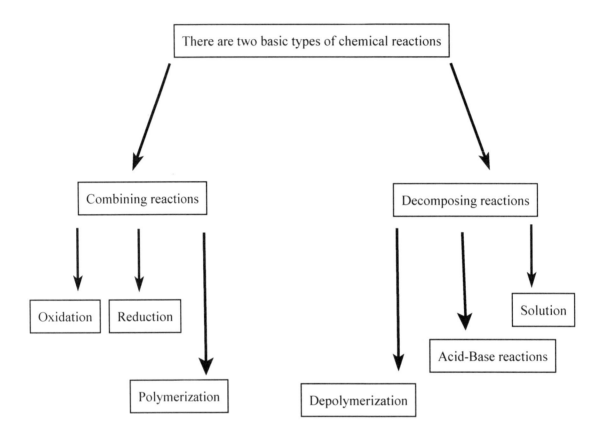

Questions for Review

Multiple-Choice Questions

1. One atom permanently receives an electron from another atom during the formation of
    a. a covalent bond
    b. an ionic bond
    c. a metallic bond
    d. a hydrogen bond

2. What type of bond is formed when two hydrogen atoms share a pair of electrons to form a hydrogen molecule?
    a. a covalent bond
    b. an ionic bond
    c. a metallic bond
    d. a hydrogen bond

3. If a sodium atom loses an electron to a chloride atom, the resulting sodium ion
    a. remains electrically neutral
    b. acquires one unit of negative charge
    c. acquires one unit of positive charge
    d. acquires two units of positive charge

4. If a chloride atom receives an electron from a sodium atom, the resulting chloride ion
    a. remains electrically neutral
    b. acquires one unit of negative charge
    c. acquires one unit of positive charge
    d. acquires two units of positive charge

5. In the periodic table (see Chapter 7), carbon is element #6. This means it has six electrons. How many valence electrons does carbon have?
    a. 1
    b. 2
    c. 3
    d. 4

6. In the periodic table, neon is element #10. How many electrons does neon have in its outermost electron shell?
    a. 2
    b. 4
    c. 6
    d. 8

7. Table salt forms a crystal in which the elements are held together by
    a.   covalent bonds
    b.   ionic bonds
    c.   metallic bonds
    d.   hydrogen bonds

8. The weakest attractive force between two or more atoms is
    a.   a covalent bond
    b.   an ionic bond
    c.   a hydrogen bond
    d.   van der Waals forces

9. During some chemical reactions, one atom transfers an electron when combining with another atom. The atom that loses the electron is said to be
    a.   oxidized
    b.   reduced
    c.   decreased
    d.   polymerized

10. Any material that releases positive hydrogen ionis when dissolved in water is
    a.   a base
    b.   an acid
    c.   a polymer
    d.   a crystal

Fill-In Questions

11. A _____ is the attractive force that holds two or more atoms together.
12. _____ electrons are involved in the formation of chemical bonds.
13. In an _____ bond, electrons are permanently transferred from one atom to another.
14. In a _____ bond, electrons are shared between two or more atoms.
15. The second electron shell out from the nucleus of the atom holds a maximum of _____ electrons.
16. Two or more metals can be combined to form an _____.
17. A molecule that has clusters of more positively and negatively charged atoms in different parts of the molecule are said to be _____.
18. The study of carbon-based molecules is called _____ chemistry.
19. In a _____, positive nuclei move about in a sea of electrons.
20. Nylon, the first human-made fiber, is an excellent example of a _____,

Crossword Quiz: The Chemical Bond

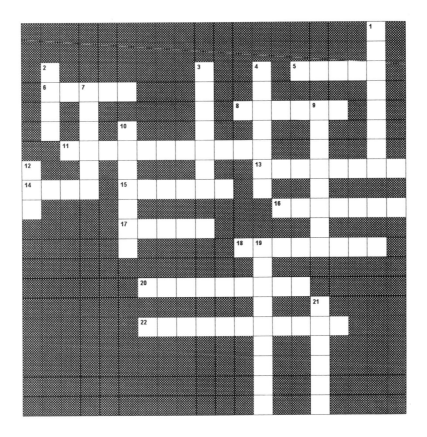

ACROSS

5. A material with a fixed shape and volume
6. A combination of two or more metals
8. Positive nuclei moving in a sea of electrons
11. The direct conversion of a solid into a gas
13. A chemical bond involving shared electrons
14. A substance that releases hydrogen ions in solution
15. A covalent bond involving only one pair of electrons
16. An extremely long molecule formed from numerous smaller molecules
17. A chemical bond created by the permanent transfer of electrons
18. Two or more atoms held together by a covalent bond
20. A chemical reaction that adds electrons to an atom
22. A molecule composed of hydrogen and carbon

DOWN

1. The simplest atom in nature
2. A substance that releases hydroxide ions in solution
3. A group of regularly repeating atoms held together by ionic bonds
4. The electrons in an atom that participate in chemical reactions
7. A material with a fixed volume but no fixed shape
9. A type of bond where electrons are shared by all of the atoms in the material
10. A solid that can be easily molded or formed
12. A substance with no fixed shape or volume
19. A chemical reaction that removes electrons from an atom
21. A covalent bond involving two pairs of electrons

Answers to Review Questions

Multiple-Choice Questions:

1. b; 2. a; 3. c; 4. b; 5. d; 6. d; 7. b; 8. d; 9. a; 10. b

Fill-In Questions:

11. chemical bond; 12. valence; 13. ionic; 14. covalent; 15. eight; 16. alloy; 17. polarized; 18. organic; 19. plasma; 20. polymer

Chapter 11

Properties of Materials

Chapter Review

The modern world is composed of both natural and man-made materials. Materials differ in their constituent atoms, the way in which those atoms are arranged, and the types of bonds that hold the atoms together. The strength of a material is related to its ability to resist crushing (compressive strength), resist pulling apart (tensile strength), and to resisting twisting (shear strength). Some materials, such as copper tubing, are made of a single element, but many important materials are composites of several elements, sometimes both natural and artificial. Some materials, primarily metals and their alloys, conduct electricity quite freely, while others have high resistance to the passage of electrons, and are used as insulators. Still other materials act as semiconductors because they carry electricity, but not very well. These materials are very useful in producing the chips that enable computers to function as they do. Superconductors lose all resistance to electricity, and conduct electricity very efficiently. Some materials display a high degree of magnetism due to the orientation of atoms within the material. Modern technology has taken advantage of many of the characteristics of materials in the development of the tools, devices and appliances that we use to save energy and labor. Diodes are semiconducting devices that permit electrons to flow in only one direction, and are capable of changing alternating current into direct current. Transistors are devices that incorporate three regions of semiconductors, and act as switches in turning current on and off. Transistors and diodes have been particularly important in the development of computers, and communications devices. Microchips are small silicon wafers onto which have been engraved numerous microscopic circuits of transistors and diodes. The computer uses these circuits to code information into a series of binary (yes-no) digits called bits. Eight bits comprise a byte; a thousand bytes comprise a kilobyte, and a million bytes comprise a megabyte, a gigabyte is one billion bytes. Even though computers are capable of performing many tasks at a very fast, and almost perfectly repeatable rate, they still are not yet as complex as the human brain, nor can they perform the same kinds of creative tasks.

Learning Objectives

After studying this chapter, you should be able to:
(Other objectives may also be assigned by your instructor)

1. Distinguish among natural, artificial, and composite materials.
2. List the features that determine the properties of a material.
3. Define the three different kinds of strength associated with a material.
4. Discuss the characteristics that determine whether a material is a conductor, insulator, or semiconductor.

5.   Discuss the concept of "conduction electrons" and "holes" relative to semiconductors.
6.   Describe how superconductors operate.
7.   Discuss how "doping" a semiconductor with different elements produces negative and positive variations of semiconductors.
8.   Describe how diodes and transistors can be used to control the flow of electrons in a circuit.
9.   Discuss how microscopic transistorized switches are used in a computer to encode and store information.

Key Chapter Concepts

- Materials available to a society often define the characteristics of that culture.
  - Stone, iron and bronze defined the sophistication of certain cultures in the past.
  - Today, we use both natural materials, such as wood, and synthetic materials, such as nylon.
  - Some materials are made by combining two or more substances and are called composite materials.
  - The properties of every material depend on three essential features:
    - the kind of atoms that make it up
    - the way those atoms are arranged
    - the way the atoms are bonded together
- Strength is the ability of a material to resist changes in its shape in three ways:
  - compressive strength resists crushing
  - tensile strength resists pulling apart
  - shear strength resists twisting
- Materials vary in their ability to conduct electricity.
  - Conductors, such as metals, are capable of carrying an electrical current.
  - Fluids, such as water and salts also conduct electricity freely.
  - Other materials have a resistance to the conductance of electricity.
  - Materials, such as wood and ceramics, have a high electrical resistance, and make good insulators.
  - Semiconductors, such as silicon, conduct electricity poorly, but enough to allow an electrical current to pass through the material.
    - The electrons that are capable of moving through a semiconductor, carrying a charge, are called conduction electrons.
    - The place in a semiconductor with a missing electron is called a hole.
    - Both conduction electrons and holes are capable of changing their positions in a semiconductor, thereby allowing an electrical current to pass through the material.
  - Superconductors exhibit very little resistance to the flow of electrons through the material.
    - Most superconductivity occurs at temperatures near absolute zero.
    - Other materials, particularly some oxides, are capable of superconductivity at temperatures as high as 160 degrees above absolute zero.
- Some materials may be treated to align their molecules in the same direction, thereby making them magnetic.
- Semiconductors, dipoles and transistors may be incorporated into electronic devices and appliances and into microscopic circuits on silicon wafers that can conduct carefully controlled electrical currents.
  - Semiconductors, such as silicon, may be "doped" by incorporating small amounts of other materials into the wafer.

- Elements like phosphorus added to the wafer provide electrons that can move as an electrical current. Such a composite is called an *n-type semiconductor* because the moving charge is a negative electron.
- Elements like aluminum absorb electrons, thereby producing holes in the composite. Such a composite is called a *p-type semiconductor* because the holes produced by missing electrons act as a positive charge.
- Diodes are semiconducting composites consisting of an n-type material bonded to a p-type material. A diode acts as a one-way gate, allowing an electrical current to pass through in only one direction.
- Semiconducting diodes are used extensively in electrical devices and appliances to direct the flow of electrical current and to change alternating current into direct current.
- In certain semiconducting diodes light pushes electrons from the n-type material into the p-type material, thereby creating an electrical current. These photovoltaic cells are thus capable of turning light energy into electrical energy.
- A transistor is a semiconducting diode in which two similar type materials are separated by the other type material, such as: n-type/p-type/n-type, or p-type/n-type/p-type.
- A transistor can be used to amplify or diminish an electrical current, and also can be used as a switch to turn a current on and off.
- Microchips are thin wafers of silicon that are engraved with diodes and transistors that are integrated into a complex electrical circuit.
- Microchips can be used to code information into binary digits called bits.
- Color television screens also utilize information in the form of bits to create a picture.
- Computers manipulate bits of electrical information in groups of eight switches (or bits). Eight bits are called a byte. The central processing unit (cpu) of the computer contains semiconductor transistors that manipulate a certain amount of material. Large amounts of information can be stored on floppy discs, or on a hard drive as magnetically oriented particles.
- Even though very complex, computers do not approach the complexity of the human brain.

Key Individuals in Science

- Karl Alex Muller and George Bednorz first discovered high-temperature superconductors in 1986.
- John Bardeen, Walter Brattain, and William Shockley invented the transistor in 1947.

Key Concept: The Characteristics of Materials

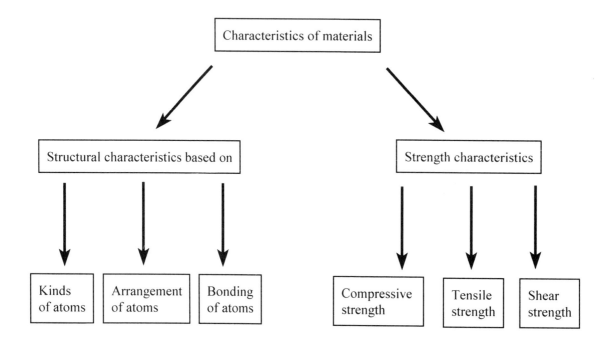

Key Concept: The Electrical Properties of Matter

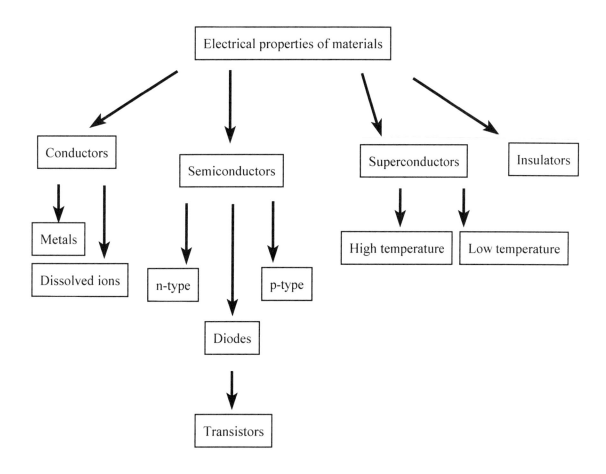

Questions for Review

Multiple-Choice Questions

1. The properties of every material depend on:
   a. the kind of atoms it is made up of
   b. the way those atoms are arranged
   c. the way the atoms are bonded together
   d. all of the above

2. Compressive strength is
   a. the ability of a material to withstand twisting
   b. the ability of a material to withstand crushing
   c. the ability of a material to withstand pulling apart
   d. none of the above

3. The reason that you are able to easily separate the particles that make up baby powder is because the molecules of baby powder are held together by
   a. covalent bonds
   b. ionic bonds
   c. hydrogen bonds
   d. van der Waals forces

4. In the strongest materials that occur in nature, the molecules are held together by
   a. covalent bonds
   b. ionic bonds
   c. hydrogen bonds
   d. van der Waals forces

5. Any material that is composed of more than one substance is called a(n)
   a. conductor
   b. composite
   c. semiconductor
   d. transistor

6. A material that conducts electricity poorly is referred to as
   a. a semiconductor
   b. a conductor
   c. an insulator
   d. a transistor

7. A material that conducts electricity with virtually no resistance at very low temperatures is
   a. a semiconductor
   b. a transistor
   c. an insulator
   d. a superconductor

8. A semiconducting device formed from one *p* and one *n* type semiconductor is
    a. a diode
    b. a transistor
    c. a superconductor
    d. a microchip

9. A photovoltaic cell can convert solar energy into
    a. movement
    b. electric energy
    c. food
    d. computer bytes

10. Eighty bits of computer information would equal how many bytes?
    a. 8
    b. 10
    c. 100
    d. 800

Fill-In Questions

11. _____ is the ability of a solid to resist changes in shape.
12. The ability to withstand pulling apart is referred to as _____ strength.
13. The point at which a material stops resisting external forces and begins to deform permanently is called its _____.
14. A diamond is one of the hardest substances know because of the numerous short _____ bonds between carbon atoms.
15. The addition of a minor impurity to an element or compound is called _____.
16. A _____ is a machine that stores and manipulates information.
17. One million bytes of computer information is also called one _____.
18. Microchips in computers are engraved with transistors and diodes in an _____ circuit.
19. The primary material used to manufacture microchips is _____.
20. A device with an n-type semiconductor and a p-type bonded to one another is a _____.

Crossword Quiz: Properties of Materials

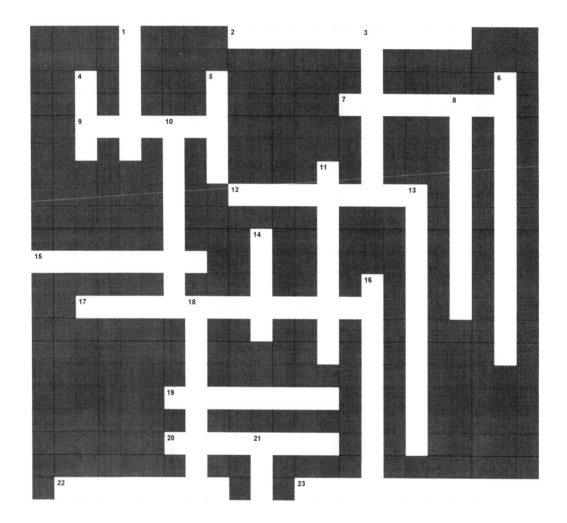

ACROSS

2. Strength that resists crushing
7. A semiconductor that contains excess electrons
9. Strength that resists pulling apart
12. A wafer of silicon used in a computer's central processing unit
15. One million bytes
17. A material that has essentially no resistance to the movement of an electric current
19. A semiconductor that has a shortage of movable electrons
20. One thousand bytes
22. A machine that stores and manipulates information
23. A material whose atoms line up into small magnetic domains

DOWN

1. The addition of minor impurities to an element or compound
3. The ability of a solid to resist changes in shape
4. Eight bits of information in a computer
5. The strength that resists twisting
6. A material that conducts electricity inefficiently
8. A circuit on a microchip consisting of transistors and diodes
10. A material that does not conduct electricity
11. A material that transmits electricity freely
13. A device that can turn light energy into electrical energy
14. A semiconductor that consists of both n-type and p-type materials
16. A device containing two of the same type of semiconductor separated by the opposite type
18. A material composed of two or more substances
21. A binary digit unit of information

Answers to Review Questions:

Multiple-Choice Questions:

1. d; 2. b; 3. d; 4. a; 5. b; 6. a; 7. d; 8. a; 9. b; 10. b

Fill-In Questions:

11. strength; 12. tensile; 13. elastic limit; 14. covalent; 15. doping; 16. computer; 17. megabyte; 18. integrated; 19. silicon; 20. diode

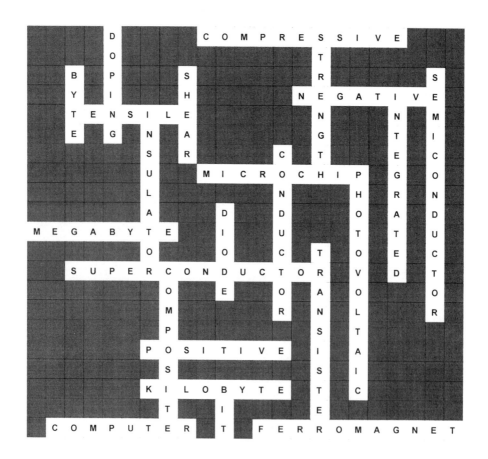

Chapter 12

The Nucleus of the Atom

Chapter Review

The central core of an atom, the nucleus, contains practically all of the mass of that atom. The nucleus consists of positively charged protons and electrically neutral neutrons. The number of protons in the nucleus determines the atomic number of the element, and the total number of protons and neutrons defines the mass number of the atom. Mass numbers may vary because different atoms of the same element contain different numbers of neutrons. Atoms of an element with different numbers of neutrons are isotopes of that element. Different isotopes display the same chemical properties because they have the same number of valence electrons (see Chapter 10). Protons and neutrons within the nucleus are held together by the strong force, a force yet incompletely understood by physicists. Some atomic nuclei undergo spontaneous decomposition in a process called radioactive decay. During alpha decay, the nucleus loses two protons and two neutrons. During beta decay, a neutron is transformed into a proton, an electron, and a neutrino. During gamma decay, the protons and neutrons within the nucleus reshuffle their energy positions, and gamma rays are emitted from the nucleus. Most nuclear radiation is dangerous, even lethal to living organisms, however, some forms have been harnessed to provide medical benefits. The half-life of an element is the time necessary for half of a mass of that element to spontaneously change to another element through radioactive decay. Knowledge of radioactive half-lives allows scientists to date anthropological and geological specimens of ancient origin. Nuclear reactors are used to employ a controlled decay reaction to generate heat, which powers a steam turbine to produce electricity. Nuclear fusion, a process in which two hydrogen nuclei combine to form a helium nucleus, with the release of much energy and radioactivity, forms the basis of the energy production in the sun.

Learning Objectives

After studying this chapter, you should be able to:
(Other objectives may also be assigned by your instructor)

1. Define the terms in Einstein's equation: $E = mc^2$.
2. Discuss the organization of the atomic nucleus, differentiation between protons and neutrons.
3. Describe how atomic number and atomic mass are related to the number of protons and neutrons in the nucleus.
4. Discuss the existence of different isotopes of elements.
5. Name the force that holds the protons and neutrons together in a nucleus.
6. Define radioactive decay, and indicate the differences among alpha, beta, and gamma decay.
7. Discuss the importance of radioactivity to human health.

8. Describe the concept of half-life, and indicate its importance to anthropologists and geologists.
9. Discuss how both nuclear fission and nuclear fusion impact on human society.

Key Chapter Concepts

- There is enormous energy contained within the nucleus of every atom.
    - What goes on in the nucleus of an atom has almost nothing to do with the atom's chemistry and vice versa.
    - The energy available in the nucleus is much greater than that available among electrons.
    - Einstein's equation defines the energy available in the nucleus: $E = mc^2$, or Energy = mass x (the speed of light)^2.
    - Even though small, the mass of any atom is multiplied by a very large number (3 x 10 meters per second, squared), producing an enormous product.
- The nucleus of the atom consists of smaller particles, protons and neutrons.
    - Each proton has a mass unit of 1, and an electrical charge of +1.
    - Each neutron has a mass unit of 1, and no electrical charge.
    - The mass of a proton or neutron is about 2000 times the mass of an electron circling the nucleus.
    - The number of protons in its nucleus determines the atomic number of an element.
    - The total mass of protons and neutrons in the nucleus determines the atomic mass of the element.
    - Some variants of elements that have the same number of protons, but different numbers of neutrons are called isotopes of that element. All isotopes exhibit the same chemical characteristics of the element because all have the same number of valence electrons.
    - Protons and neutrons are held together in the nucleus by the strong force, a very powerful force acting over a very short distance.
- A very small percentage of atomic nuclei are unstable; they spontaneously emit particles and energy and are transformed into different elements.
    - The spontaneous decomposition of an atomic nucleus releases particles and energy and is called radioactive decay or radioactivity.
    - Alpha decay involves the emission of two protons and two neutrons from a decomposing nucleus. These particles may combine to form a helium atom.
    - Beta decay involves the emission of an electron from a decomposing nucleus.
    - Gamma decay involves the reshuffling of protons and neutrons with respect to their energy levels, and releases gamma-rays or X-rays. The element retains its identity.
    - Radioactive decay is usually very harmful, even lethal, to living organisms. Nuclear explosions and failures of nuclear reactors have resulted in releases of radioactivity and energy that has killed and injured many people.
    - Certain radioactive materials are used by medical scientists to track and control certain diseases.
    - The half-life of an element is the time it takes for half of a mass of radioactive material to undergo decay to another element.
    - The concept of half-life leads to radiometric dating, the process of determining how long a certain amount of radioactive material has been in a certain place. Anthropologists and geologists use the technique daily to relate the age of specimens they collect in the field.
    - Every radioactive element has a characteristic decay chain that leads from the original nucleus to its eventual decay product.

- Radon, an odorless, colorless, inert, radioactive gas may seep into closed basements and buildings, and is dangerous to human life.
- When an atomic nucleus decomposes, large amounts of energy in the form of radioactivity and heat are released.
  - Humans have learned to use the heat released during fission to produce electricity.
  - Nuclear reactors use heat given off from the fission reaction in uranium to drive a steam turbine that powers an electric generator.
  - Numerous safety features are necessary to prevent the meltdown of the reactor and the release of radioactive materials into the surrounding environment.
  - Nuclear fusion of the combination of two nuclei of low atomic numbers to produce a nucleus with a higher atomic number.
  - During the process of fusion, enormous quantities of energy in the form of radiation and heat are released. The fusion of hydrogen nuclei to produce helium is the basis for the heat and light that emanate from the sun.
  - Fusion is most often witnessed on earth during the explosion of a thermonuclear (hydrogen) weapons. Relatively small amounts of certain isotopes of hydrogen can be used to produce devastating destructive effects.

Key Individuals in Science

- Antoine Becquerel (1852-1908) originally discovered the existence of radioactivity.
- Marie Curie (1867-1934) isolated radium, polonium and other radioactive elements.
- Ernest Rutherford (1871-1937) discovered alpha decay in radioactive uranium-238.
- Wolfgang Pauli postulated the existence of the neutrino, a tiny particle with no electric charge and no mass.

Key Concept: Radioactive Decay

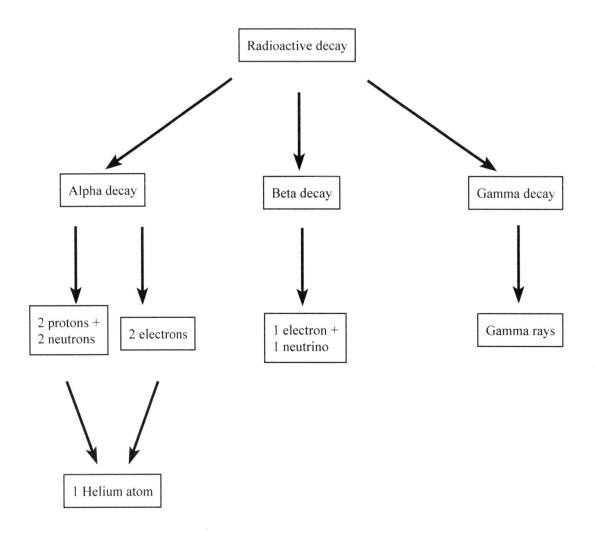

Key Concept: Beneficial Uses of Radioactivity

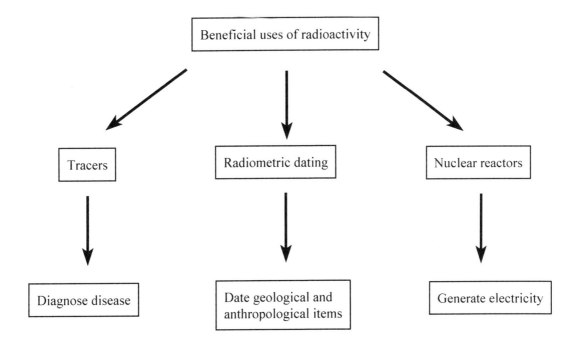

Questions for Review

Multiple-Choice Questions

1. When the energy available within the nucleus is released, it follows
   a. Newton's laws of gravity
   b. Einstein's law relating energy to mass and the speed of light
   c. Rutherford's rule of radioactive decay
   d. none of the above

2. On the average, each person in the United States uses energy at the rate of about
   a. one kilowatt-hour/each hour
   b. ten kilowatt-hours/each hour
   c. 100 kilowatt-hours/each hour
   d. 1000 kilowatt-hours/each hour

3. The positively-charged particle found within the nucleus of the atom is
   a. the electron
   b. the neutron
   c. the proton
   d. the neutrino

4, The tiny particle released during beta decay that has no electric charge is
   a. an electron
   b. a neutrino
   c. a proton
   d. a neutron

5. During Rutherford's experiment documenting alpha decay, alpha particles led to the production of
   a. hydrogen gas
   b. neutrinos
   c. gamma rays
   d. helium atoms

6. The type of radioactive decay that causes the release of two protons and two neutrons is
   a. alpha decay
   b. beta decay
   c. gamma decay
   d. none of the above

7. Radioactivity was first discovered by
   a. Marie Curie
   b. Antoine Becquerel
   c. Pierre Curie
   d. Albert Einstein

8. During the nuclear fission reaction in a reactor that generates electricity,
   a. two alpha particles unite to form a molecule of helium
   b. uranium-238 releases beta particles
   c. uranium-235 releases neutrons
   d. uranium-234 atoms fuse, with the release of large amounts of energy

9. The half-life of a radioactive element
   a. is the time it takes for half of the atoms to spontaneously decay
   b. indicates that it is capable of reducing the life of a human exposed to it by one-half
   c. describes half the energy that is released with the atomic nucleus decays
   d. is a measure of half the radioactivity that emanates from each atomic nucleus that decays

10. Radioactive tracers are used
    a. to date geological and anthropological items
    b. to produce electrical energy in a fission reactor
    c. to produce electrical energy in a fusion reactor
    d. to help diagnose disease

Fill-In Questions

11. The quantity of uranium-235 necessary to produce an explosive chain reaction is called the
    _____.

12. Unlike the fission bomb, the hydrogen bomb relies on nuclear _____.

13. In the equation $E = mc^2$, the symbol "c" stands for the _____.

14. The number of protons in the nucleus is called the atom's atomic _____.

15. A _____ is a form of an atom that has the same number of protons but different numbers of neutrons.

16. The atomic number of the element oxygen is 8. Oxygen 18 must have _____ neutrons in its nucleus.

17. The force that holds protons and neutrons together in the nucleus of the atom is called the _____ force.

18. Beta radioactive decay results in the release of _____ from the decaying atom.

19. A _____ is a radioactive element that can be injected into a person to detect the presence of diseased cells that will accumulate the element.

20. The splitting of an atomic nucleus is called fission, whereas the combination of two atomic nuclei is referred to as _____.

Crossword Quiz: The Nucleus of the Atom

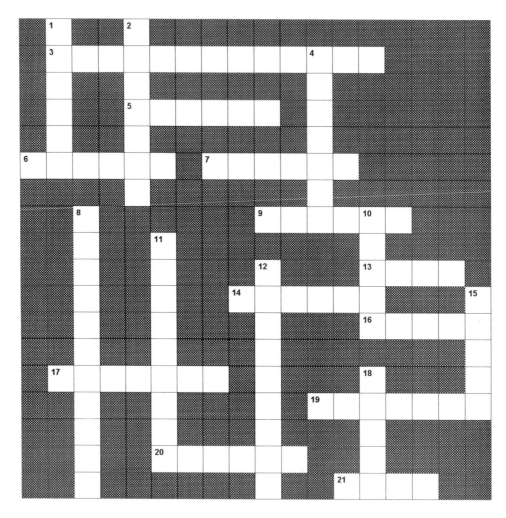

ACROSS

3. The ability of certain atoms to spontaneously decay and emit energy and subatomic particles
5. This force holds the atomic nucleus together
6. The positively charged particle found in the nucleus of the atom
7. The combination of two nuclei to form a new element
9. During nuclear fission, mass is converted into _____
13. The total number of protons and neutrons in the nucleus of an atom is called the _____ number
14. The element formed during alpha decay
16. This type of radioactive decay releases pairs of protons and neutrons
17. A device that uses nuclear energy to produce electricity
19. The electrically neutral subatomic particle
20. The total number of protons in the nucleus is called the atomic _____
21. This type of radioactive decay emits electrons

DOWN

1. A radioactive material used to locate diseased cells in the body
2. The splitting apart of an atomic nucleus
4. A form of an element having different numbers of neutrons in the nucleus
8. The technique using radioactive materials to determine the age of an object is called _____ dating
10. The type of radioactive decay that emits high-energy photons
11. Alpha, beta, and gamma decay are all forms of _____
12. The scientist who discovered radiation
15. This radioactive gas may accumulate in the basement of a house
18. The scientist who isolated radium and polonium

Answers to Review Questions:

Multiple-Choice Questions:

1. b; 2. a; 3. c; 4. b; 5. d; 6. a; 7. b; 8. c; 9. a; 10. d

Fill-In Questions:

11. critical mass; 12. fusion; 13. the speed of light; 14. number; 15. isotope; 16. ten; 17. strong; 18. electrons; 19. tracer; 20. fusion

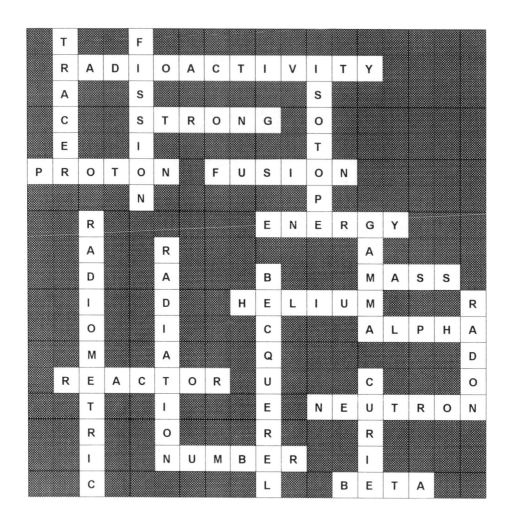

Chapter 13

The Ultimate Structure of Matter

Chapter Review

Physicists have discovered what appear to be the ultimate building blocks of matter. The elements found in the universe are composed of discrete units called atoms. Atoms, in turn, are composed of smaller particles. Hadrons are the nuclear particles, and include the proton and neutron. Leptons are found outside the nucleus, and are the electron and neutrino. Hadrons and leptons also have antimatter particles, such as antiprotons and antineutrons. Hadrons are composed of still smaller particles called quarks. Quarks differ from other particles in that they have fractional ($+\frac{1}{3}$, $-\frac{1}{3}$, $+\frac{2}{3}$, $-\frac{2}{3}$) electrical charges, and cannot exist alone in nature. Quarks combine in different proportions to form all of the larger particles. Scientists have yet to see quarks in the laboratory, yet their theoretical existence makes it possible to explain the phenomena associated with nuclear events. There are four forces in nature. The strong force holds the hadrons together in the nucleus. The weak force is responsible for radioactive decay. The forces of gravity and electromagnetism are part of our everyday experience, and operate throughout the universe. Scientists are currently attempting to discover a unified theory that explains the four fundamental forces as different aspects of a single unified force.

Learning Objectives

After studying this chapter, you should be able to:
(Other objectives may also be assigned by your instructor)

1. Describe a particle accelerator, and tell how it is used to study the contents of the nucleus.
2. Distinguish among hadrons, leptons, and antiparticles.
3. Define antimatter, and indicate how it is produced in the laboratory.
4. Indicate how positrons are used in medical diagnosis.
5. Enumerate the different kinds of quarks found within the nucleus.
6. Discuss why quarks cannot exist alone in nature.
7. Describe the four fundamental forces, and tell how they differ from one another.
8. Discuss the concept of a unified field theory, and discuss the importance of such a theory.

Key Chapter Concepts

- Matter in the universe is organized at several different levels.
  - Elements are composed of atoms of the same kind.
  - Atoms have a nucleus containing protons and neutrons

- Electrons orbit the nucleus.
- Scientists began studying the makeup of protons, neutrons, and electrons using cosmic rays (mostly protons) from outer space to bombard a target element.
- By studying the particles emitted from the collisions, they discovered that the nucleus held many more than just two elementary particles.
- Since that time, scientists have built devices, called particle accelerators, that use large magnets and electromagnetic radiation to impart very high velocity and energy to electrons and other subatomic particles.
  - The synchrotron is a circular device that causes charged particles to move in a circular path, while their velocity and energy are increased.
  - The linear accelerator moves particles in a straight line while increasing their velocity and energy.
  - To date, more than 200 different "elementary particles" have been identified by studying what is emitted when these particles collide with one another, or with a target.
- At the present time, scientists classify the elementary particles into three basic types.
  - Leptons are non-nuclear elementary particles that include electrons, neutrinos, and mu and tau particles. Leptons do not participate in the strong forces that hold the nucleus together.
  - Hadrons are a large variety of nuclear particles, including the proton and neutron. Hadrons participate in the strong forces that hold the nucleus together.
  - Antimatter includes an antiparticle for every particle that has been discovered in the universe. Antiparticles have the same mass as their matter twins, but have opposite charges, and opposite magnetic characteristics. When matter and antimatter particles collide, their masses are converted completely to energy in a process called annihilation.
  - Antimatter, such as the positron (a positive version of an electron), are useful in medical research to detect certain types of cellular activity in internal body organs.
- Quarks appear to be the truly fundamental building blocks of hadrons.
  - Quarks have mass and fractional electrical charges ($\pm\frac{1}{3}$ or $\pm\frac{2}{3}$ the charge of an electron or proton).
  - Six types of quarks exist: down, up, strange, charm, bottom, and top.
  - Different combinations of quarks produce all of the hundreds of elementary particles that have been discovered in the nucleus.
  - Once locked within a particle, the quarks cannot be released as free, independent units.
  - Quarks probably existed as independent units only at the very beginning of the universe.
- Four fundamental forces act to hold and arrange different parts of the universe.
  - Gravity (see Chapter 2) is a force that has infinite range, and operates on all objects in the universe.
  - The electromagnetic force also operates over long distances, and is familiar in everyday experiences.
  - The strong force operates within the nucleus of the atom, and is responsible for holding the hadrons within the nucleus.
  - The weak force is responsible for processes such as beta decay (see Chapter 11) that disrupt nuclei and destroy elementary particles.
  - Each force is activated as two objects exchange gauge particles. The graviton is the gauge particle of gravity; the photon is the gauge particle of electromagnetism; the gluon is the gauge particle of the strong force; and the W and Z particles are the gauge particles of the weak force.

- Scientists are trying to link the four fundamental forces in a unified field theory, in which the fundamental forces are seen as different aspects of the same force.
- Scientists are also trying to develop Theories of Everything (TOEs) to explain the existence of quarks, leptons, and the equivalence of all the forces in a single set of equations, and Theories of Organization (TOOs) that reveal how matter and energy are ordered into larger, complex systems.

Key Individuals in Science

- Ernest O. Lawrence produced the cyclotron, the first particle accelerator, in the early 1930s.
- Carl Anderson first discovered proof of the existence of antimatter in 1932.
- Richard Feynman (1918-1988) provided a model for understanding particle interactions and the fundamental forces.
- Stephen W. Hawking introduced the concept of a Theory of Everything (TOE), a theory to relate elementary particles and fundamental forces in one series of equations.

Key Concept: Elementary Particles

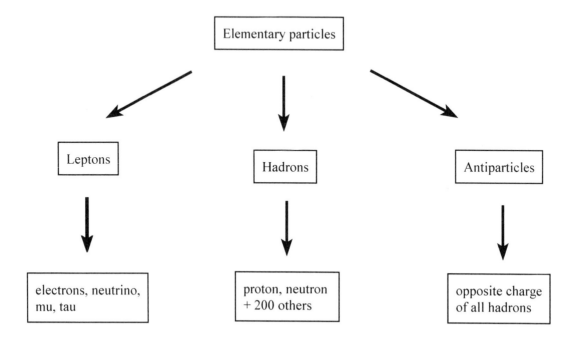

Key Concept: Fundamental Forces

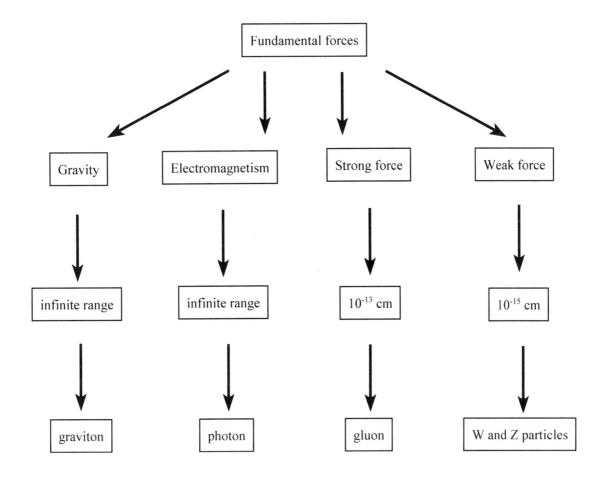

Questions for Review

Multiple-Choice Questions

1. The quest for the ultimate building blocks of the universe is referred to by philosophers as
   a. the scientific method
   b. reductionism
   c. high-energy physics
   d. an impossible task

2. Which of the following is <u>not</u> classified as an elementary particle?
   a. proton
   b. neutron
   c. atom
   d. electron

3. The elementary particle with the mass of an electron but with a positive charge is the
   a. proton
   b. neutron
   c. gamma ray
   d. positron

4. All hadrons
   a. are found in the nucleus
   b. are found outside of the nucleus
   c. have a positive charge
   d. have a negative charge

5. All leptons
   a. are found in the nucleus
   b. do not take part in the strong force
   c. take part in the strong force
   d. participate in holding the nucleus together

6. The physicist who discovered antimatter was
   a. Richard Feynman
   b. Carl Anderson
   c. Ernest Lawrence
   d. Stephen Hawking

7. The fundamental building blocks of "elementary particles" are called
   a. quarks
   b. leptons
   c. hadrons
   d. positrons

8. The fundamental force that operates over great distances is
   a. strong force
   b. weak force
   c. gravity
   d. none of the above

9. The photon is the gauge particle of
   a. gravity
   b. electromagnetism
   c. the strong force
   d. the weak force

10. The gluon is the gauge particle of
    a. gravity
    b. electromagnetism
    c. the strong force
    d. the weak force

Fill-In Questions

11. _____ are particles (mostly protons) that rain down continuously on the atmosphere of the Earth after being emitted by stars in our galaxy and in others.
12. The _____ is a circular type of particle accelerator.
13. A _____ particle accelerator moves particles in a straight line.
14. Non-nuclear elementary particles are called _____.
15. The antimatter equivalent of the proton is the _____.
16. Hadrons are found in the _____.
17. Hadrons are composed of _____.
18. The gauge particle for the electromagnetic force is the _____.
19. The strong force has its effect in the _____ of the atom.
20. The gauge particle for the strong force is the _____.

Crossword Quiz: The Ultimate Structure of Matter

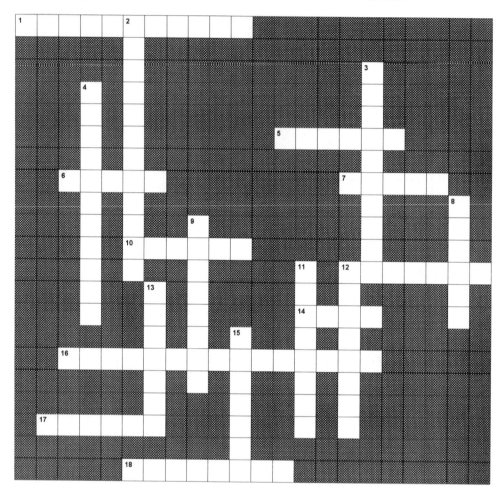

ACROSS

1. A circular particle accelerator
5. A type of high-energy ray that originates in outer space
6. The gauge particle of the strong force
7. The particle that is exchanged between two objects with one of the four fundamental forces
10. The force that holds the nucleus together
12. The fundamental force that keeps the earth orbiting the sun
14. The fundamental force that is active in beta decay
16. The fundamental force that uses the photon
17. The six most elementary particles
18. The discoverer of antimatter

DOWN

2. The search for the ultimate building blocks of the universe
3. Particles with the same mass and opposite charge of the elementary particles
4. A device used to increase the velocity and energy of particles
8. The gauge particle of the electromagnetic force
9. The antimatter representative of the electron
11. He invented particle accelerators
12. The gauge particle of gravity
13. A neutrino is an example of this type of elementary particle
15. A proton is an example of this type of elementary particle

Answers to Review Questions

Multiple-Choice Questions:

1. b; 2. c; 3. d; 4. a; 5. b; 6. b; 7. a; 8. c; 9. b; 10. c

Fill-In Questions:

11. cosmic rays; 12. synchrotron; 13. linear; 14. leptons; 15. antiproton; 16. nucleus; 17. quarks; 18. photon; 19. nucleus; 20. gluon

Chapter 14

The Stars

Chapter Review

Heavenly objects include planets, stars, and galaxies. Astronomy is the study of these heavenly bodies. Stars form from the condensation of matter in immense clouds of interstellar dust. As matter accumulates in a central region of the cloud, the gravitational pressure causes the fusion of atoms, mostly hydrogen gas, forming helium and releasing enormous amounts of light and energy. In our Sun, an example of an average star, the process takes place in three stages. First, two protons fuse to form deuterium, an isotope of hydrogen. Next, another proton fuses with the deuterium to form helium-3. Finally, two helium-3 nuclei fuse to form helium-4 plus protons and photons. In addition to heat and light, the Sun emits an outward flow of particles, the solar wind that affects all of the planets in our solar system. Our Sun, and other stars that use hydrogen as their energy source are called main-sequence stars. Some stars are white dwarfs that are very small and very hot, while others are red giants and have cool surfaces. We use both earth-based and orbiting telescopes to study stars. Some telescopes detect energy outside of the visible spectrum, such as X-rays, infrared radiation, microwave radiation, and gamma rays. The distances between stars are measured in light-years - the distance light travels in one year. All stars pass through a life cycle that ends with their death. In about six billion years, the sun will run out of hydrogen to burn. The helium remaining will be converted to carbon, which will not undergo nuclear reactions and the Sun will first collapse, and they expand into a red giant, finally shrinking into a white dwarf. Very large stars with more than 10 times the amount of hydrogen present in the Sun suffer a violent death. These stars generate sufficient temperature and pressure to cause the carbon formed in their interior to undergo nuclear reactions. This leads to a catastrophic explosion called a supernova. A very dense neutron star may result along with a supernova. Black holes may result from the collapse of the largest stars. Black holes are so massive they do not even allow light to escape from their gravitational pull. Recent observations have revealed what may be planetary systems other than our own.

Learning Objectives

After studying this chapter, you should be able to:
(Other objectives may also be assigned by your instructor)

1. Distinguish between a star and a planet.
2. Calculate the amount of energy given off by the sun.
3. Describe the process by which the sun generates energy.
4. Describe the internal structure of the sun.
5. Define the solar wind, and indicate its effects on the planets.

6. Describe several types of earth-based and orbiting telescopes.
7. Define the term "light-year," and indicate its importance in the astronomical distance scale.
8. Describe the three major types of stars.
9. Discuss the death of the Sun and other larger stars.
10 Describe neutron stars and black holes.

Key Chapter Concepts

- Astronomy is the study of heavenly objects.
- Astronomers use telescopes to study the stars.
- Most telescopes are earth-based.
  - Optical telescopes detect visible light coming from heavenly bodies.
  - Radio telescopes detect waves in other parts of the electromagnetic spectrum emitted from stars.
- Orbiting telescopes are able to avoid the interference of the atmosphere that earth-based telescopes must endure.
  - The Hubble Space Telescope (HST) measures both visible and ultraviolet radiation. It is used to observe both planetary and stellar phenomena.
  - The Space Infrared Telescope Facility (SIRTF) surveys the infrared radiation coming to the Earth from objects in space.
  - The Chandra X-Ray Observatory produces high-resolution X-ray images of objects in the sky.
- The sun has a definite internal structure.
  - The central core of the sun is where the energy-producing reactions occur.
    - Hydrogen is converted into helium, and protons and positrons collide. Mass lost in these processes is turned into energy and radiated outward.
    - From the central core outward, collisions of subatomic particles produce more energy.
  - The convection zone begins four-fifths of the way to the surface. Convection in the hydrogen-rich material in this region brings energy to the surface.
  - The relatively thin (about 150 kilometers thick) outer layer of the sun is called the photosphere. This region radiates most of the light we see from the sun and the heat we feel.
  - The outermost halo of the photosphere, the chromosphere and corona are only visible during a total eclipse of the sun.
  - The solar wind is a constant stream of particles (mostly ions of hydrogen and helium) from the photosphere into the surrounding space.
- The Sun is a typical, average star.
- Energy is produced in the Sun via a fusion process.
  - A star, such as the Sun, is born when matter in an interstellar cloud accumulates in one region.
  - Gravity causes a very large mass of matter, mostly hydrogen gas, to concentrate, increasing the temperature and pressure at the center of the mass.
  - As temperature and pressure rise, a plasma (see Chapter 9) of protons develops, and they start to fuse. Energy is produced in a three-step process as mass is converted into energy:
    - Two protons fuse to form deuterium (an isotope of hydrogen made up of one proton and one neutron), a positron (see Chapter 12) and a neutrino.
    - Another proton fuses with the deuterium to form helium-3 with two photons and one neutron. Gamma rays are also produced.
    - Two helium-3 nuclei fuse to form helium-4, two protons and another gamma ray.

- Stars differ in their luminosity.
  - The luminosity is the total energy emitted by a star.
  - The apparent magnitude of a star is its brightness when viewed from Earth.
  - The absolute magnitude of a star is its brightness if viewed at a standard distance.
- Stars differ in their distance from earth.
  - The distance of stars from earth is usually measured in light-years, the distance light travels in one year, about 10 trillion kilometers (about 6.2 trillion miles).
  - The distance to nearby stars can be measured by triangulation.
  - The distance to a Cepheid variable star is determined by measuring the regular dimming-brightening-dimming behavior that these stars exhibit.
- Stars can be grouped into three major types using the Hertzsprung-Russell diagram.
  - Main-sequence stars are in the hydrogen-burning phase of their lives, and their energy is derived from the hydrogen to helium fusion reactions already described.
  - Red giants are very large stars that emit a lot of energy but their surfaces are very cool.
  - White dwarfs are small, very hot stars that emit little energy.
- Stars pass through a predictable life cycle.
  - A star is born when the gas and dust in a nebula collapses on itself under the influence of gravity.
  - The big lump of material attracts material from surrounding regions, and once it reaches a critical size, the internal pressure and temperature become high enough to initiate nuclear fusion reactions.
  - Average hydrogen-burning stars, such as the Sun, will eventually run out of their hydrogen fuel, and begin to collapse.
  - As the Sun collapses, the remaining helium will fuse to form carbon-12, in a process called helium burning. Carbon will not participate in nuclear reactions in the Sun.
  - After the collapse, the Sun will become a red giant, ballooning out far enough to swallow the planet Mercury.
  - Following the red giant stage, the Sun will collapse into a white dwarf.
  - Very large stars may generate sufficient heat and pressure during their collapse that carbon that accumulates at the core will participate in violent nuclear reactions producing a layered internal structure.
  - Eventually this large star will collapse, triggering a violent explosion called a supernova.
  - The core of neutrons left from the supernova explosion forms a neutron star.
  - The neutron star develops a strong magnetic field, and emits large quantities of charged particles into space.
  - Neutron stars often spin very rapidly, producing regular magnetic pulses. These stars are now called pulsars.
  - Occasionally, a large star collapses completely producing a black hole. A black hole is an object so massive that it does allow anything, not even light, to escape from its gravitational field.

Key Individuals in Science

- Pierre Laplace (1749-1827) first proposed the nebular hypothesis model of star formation.
- Freeman Dyson speculated about future civilizations constructing a "Dyson sphere" around a star, thus absorbing and using all of the energy emitted by the star.
- Ejnar Hertzsprung and Henry Russell developed the Hertzsprung-Russell diagram to classify stars based on their luminosity and surface temperature.

Key Concept: The Life and Death of the Sun

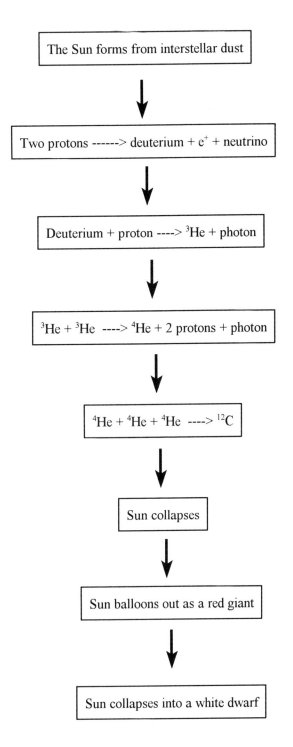

The Sun forms from interstellar dust

↓

Two protons ------> deuterium + e$^+$ + neutrino

↓

Deuterium + proton ----> ^{3}He + photon

↓

^{3}He + ^{3}He ----> ^{4}He + 2 protons + photon

↓

^{4}He + ^{4}He + ^{4}He ----> ^{12}C

↓

Sun collapses

↓

Sun balloons out as a red giant

↓

Sun collapses into a white dwarf

Questions for Review

Multiple-Choice Questions

1. During the first step of energy production in the Sun,
   a. two helium-3 nuclei fuse to form deuterium and a neutrino
   b. deuterium and a proton fuse to form hydrogen atoms
   c. two protons fuse to form deuterium, a positron and a neutrino
   d. three helium-3 nuclei fuse to form a carbon-12 nucleus

2. The net effect of the three-step process that produces energy in the sun is called
   a. hydrogen burning
   b. helium burning
   c. deuterium burning
   d. proton burning

3. The most abundant element in the universe is
   a. helium
   b. carbon
   c. hydrogen
   d. neutrinos

4. The region of the sun's interior that conducts energy between the innermost portion and the outermost portion is
   a. the chromosphere
   b. the convection zone
   c. the corona
   d. the photosphere

5. The sun's output of energy is in the _____ of the visible spectrum.
   a. middle
   b. shortest wavelength area
   c. longest wavelength area
   d. it is distributed evenly through the electromagnetic spectrum

6. Which of the following is an earth-based telescope?
   a. Hubble telescope
   b. Gamma Ray Observatory
   c. Cosmic Background Explorer
   d. Keck telescope

7. A light-year is
   a. the distance the Space Shuttle can travel in one year
   b. the distance that light travels in one year
   c. the time it takes for light to travel from the Sun to the earth
   d. none of the above

8.   The Sun is currently classified as
     a.   a white dwarf
     b.   a red giant
     c.   a white giant
     d.   a main-sequence star

9.   The Hertzsprung-Russell diagram is used to
     a.   classify stars based on luminosity and surface temperature
     b.   determine the distance of a star from the earth
     c.   determine the type of radio waves being emitted from a star
     d.   triangulate the distance from the earth to nearby stars.

10. The catastrophic explosion that occurs in some very large stars near the end of their lives produces
     a.   a neutron star
     b.   a white dwarf
     c.   a red giant
     d.   a white giant

Fill-In Questions

11. A _____ is a star that rotates rapidly and emits regular bursts of electromagnetic energy.
12. A very massive object in the universe that will not allow the escape of anything from its gravitational field is a _____.
13. The interior of a very large star that is near death is composed of _____.
14. The element that builds up in the interior of large stars that does not release energy by any kind of nuclear reaction is _____.
15. The sun will run out of hydrogen to burn approximately _____ years from now.
16. During helium burning, helium nuclei fuse to produce the element _____.
17. A small star that has a very high surface temperature, but emits low amounts of energy is a _____.
18. The process that utilizes precise sightings of the same star from opposite ends of Earth's orbit is called _____.
19. The brightness of a star as it appears from earth is called its _____ magnitude.
20. The stream of charged particles that flows from the sun is called the _____.

Crossword Quiz: The Stars

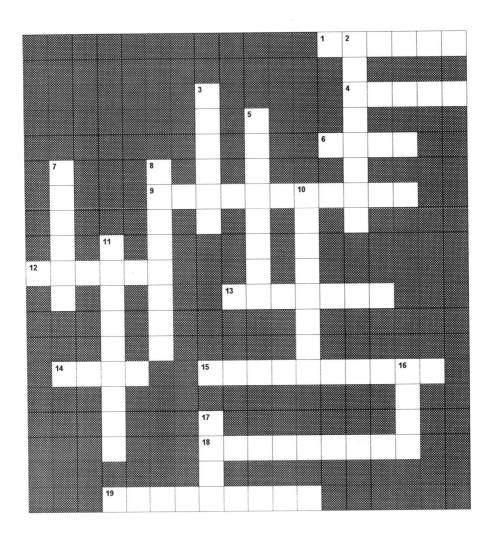

ACROSS

1. A type of gamma ray produced during energy production in the Sun
4. This scientist proposed tapping the entire energy output of a star
6. The central portion of a star that produces energy through nuclear reactions
9. The outer layer of the Sun that radiates energy into space
12. A neutron star that emits regular bursts of electromagnetic energy
13. A type of star that results from the collapse of a large star in a supernova
14. Solar _____, the stream of charged particles that flows from the Sun
15. The pressure that prevents electrons from occupying the same space in a collapsing star
18. A device used to study the stars
19. A catastrophic explosion of a very large star as it dies

DOWN

2. The most abundant element in the universe
3. The nuclear reaction that produces energy within the Sun
5. The magnitude of the luminosity of a star when viewed from a standard distance
7. The element formed from the fusion of hydrogen nuclei in the core of the Sun
8. The magnitude of the luminosity of a star when viewed from the Earth
10. A positively charged electron
11. The study of the stars
16. The orbiting telescope that monitors background microwave radiation in the universe
17. An immense fusion reactor in space

Answers to Review Questions

Multiple-Choice Questions

1. c; 2. a; 3. c; 4. b; 5. a; 6. d; 7. b; 8. d; 9. a; 10. a

Fill-In Questions

11. pulsar; 12. black hole; 13. iron; 14. iron; 15. 6.5 billion; 16. carbon; 17. white dwarf; 18. triangulation; 19. apparent; 20. solar wind

Chapter 15

Cosmology

Chapter Review

Cosmology is the study of the universe. One of the most influential pioneers of this science was Edwin Hubble, who discovered that the universe is composed of countless galaxies, in addition to the Milky Way. He also devised a way to determine the distance to remote galaxies by measuring the amount of redshift in light coming from them. His observations indicated that the farther away a galaxy was, the faster it was moving away, which led to the concept of an expanding universe. His studies also led to the realization that there were several types of galaxies, based on size, shape, and the kind of radiation they emit. The most accepted theory to explain the expanding universe is the big bang theory. This theory states that the universe began as a single event, an enormous explosion of a single, small, dense collection of matter. Evidence for the big bang includes Hubble's observations, analysis of the background microwave radiation from outer space, and the abundance of light elements, primarily hydrogen, in the universe. According to the big bang theory, all matter and forces were incorporated into a singular entity of unimaginable density and temperature. The big bang launched a series of six "freezings. " At 10^{-43} second, 10^{-35} second, and 10^{-10} second, the single unified force split into the four forces we recognize today: gravity, strong, electromagnetic, and weak. Matter existed only in the form of quarks and leptons at this early stage of the universe. During this early time, practically all antimatter was eliminated, and the universe experienced a tremendously rapid expansion. At 10^{-5} second protons and neutrons formed as quarks fused together. At 3 minutes, atomic nuclei formed, and at 500,000 years, electrons joined with the nuclei to form atoms. Today, astronomers theorize that fully 90% of the matter present in the universe is undetectable using instruments presently available. This invisible material is often referred to as "dark matter."

Learning Objectives

After studying this chapter, you should be able to:
(Other objectives may also be assigned by your instructor)

1. Distinguish between a star and a galaxy, and indicate the major types of galaxies present in the universe.
2. Discuss the contributions made by Edwin Hubble to the study of cosmology.
3. Describe the redshift phenomenon and indicate its importance in estimating the distance of galaxies from the Earth.
4. Discuss the big bang theory of the origin of the universe, and indicate several lines of evidence for this theory.
5. Describe the general characteristics of an expanding universe.
6. Discuss the six "freezings" that occurred in the early life of the universe.

7.   Describe the concept of "dark matter," and indicate its importance in the universe.

Key Chapter Concepts

- A galaxy is a large collection of millions to hundreds of billions of stars, together with gas, dust, and other materials that is held together by the forces of mutual gravitational attraction.
- There are several types of galaxies.
  - About 75% of the brightest galaxies are spiral galaxies, like the Milky Way - flattened disks with a central bulge, and bright spiral arms stretching out from the center.
  - Elliptical galaxies are large football-shaped collections of stars that make up about 20% of the brightest galaxies.
  - Small irregular and dwarf galaxies make up the remainder of galaxies.
  - Quasars are wild, explosive, distant objects in the universe that emit tremendous amounts of energy.
- The farther away from the Earth a galaxy is, the faster it is traveling .
  - This phenomenon was discovered by studying the light emitted by galaxies.
  - Light emitted from receding galaxies is redshifted relative to its distance from the Earth and its speed of travel.
- The big bang theory states that the universe began at a specific point in the past, and has been expanding ever since.
  - Today, galaxies are clumped together into groups and clusters, many of which are, in turn, grouped into larger collections called superclusters of thousands of galaxies surrounding voids (empty places) in the universe.
  - This movement can be visualized as raisins in a loaf of rising dough in a bakery, or as dots on an expanding balloon, all moving away from one another as the dough rises or balloon expands.
  - Several lines of evidence have been discovered that support the big bang theory.
    - The universal expansion discovered by Edwin Hubble supports the theory.
    - The discovery of cosmic microwave background radiation coming from all parts of the universe supports the concept of the continued cooling of a universe that began in an incredibly monumental explosion at some time in the past.
    - The abundance of light elements supports this theory. Only light nuclei: hydrogen, helium, and lithium, were able to form during the early stages of the universe. Heavy elements are formed later in stars and supernovae (see Chapter 14).
- The early stages of the universe included six different phenomena, often called "freezings," that occurred at specific times:
  - At 10^{-43} second following the big bang, the force of gravity separated from the strong-electroweak force.
  - At 10^{-35} second following the big bang, the strong force separated from the electroweak force.
  - At 10^{-10} second following the big bang, the weak and electromagnetic forces separated.
    - At this time, practically all antimatter was annihilated.
    - A short, rapid, expansion (inflation) of the universe occurred at this time.
  - At 10^{-5} second following the big bang, quarks fused to form hadrons and leptons.
  - At three minutes following the big bang, atomic nuclei formed.
  - At 500,000 years following the big bang, electrons became associated with atomic nuclei to form complete atoms.

- The luminous objects in the sky constitute only about 10% of the matter in the universe. Dark matter that cannot be detected using present-day instruments comprises approximately 90% of the matter in the universe.

Key Individuals in Science

- Edwin Hubble (1889-1953) made several important cosmological discoveries.
  - Using the 100 inch telescope on Mount Wilson, California, he measured the distance to Cepheid variable outside of the Milky Way, thereby establishing the existence of galaxies outside of the Milky Way.
  - He discovered that the distinctive colors emitted by different elements in other galaxies were shifted toward the red (long-wavelength) end of the spectrum, compared to light emitted by atoms on Earth. This redshift was an example of the Doppler effect, indicating that the galaxy was moving away from the Earth.
  - He stated Hubble's law: the farther away a galaxy is, the faster it recedes.
- Arno Penzias and Robert W. Wilson discovered cosmic microwave background radiation in 1964.
- Margaret Geller and John Huchra discovered that galaxies were distributed throughout the universe in clusters surrounding empty voids, like soapsuds.

Key Formulas and Equations

- Hubble's law: a galaxy's velocity = (Hubble constant) x (galaxy's distance), or $v = H x d$

Key Concept: The Evolution of the Universe

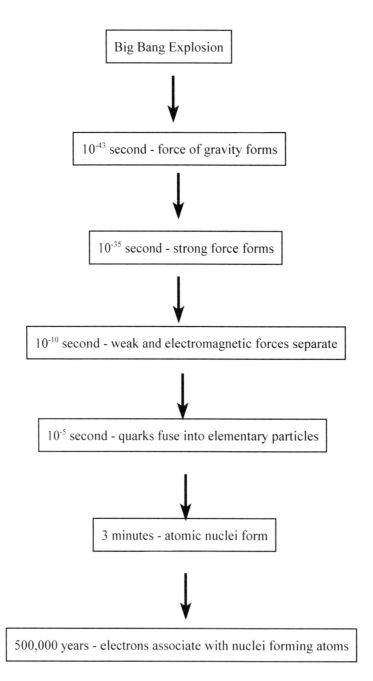

Big Bang Explosion

↓

10^{-43} second - force of gravity forms

↓

10^{-35} second - strong force forms

↓

10^{-10} second - weak and electromagnetic forces separate

↓

10^{-5} second - quarks fuse into elementary particles

↓

3 minutes - atomic nuclei form

↓

500,000 years - electrons associate with nuclei forming atoms

Questions for Review

Multiple-Choice Questions

1. A large assembly of stars together with gas, dust, and other materials that is held together by the forces of mutual gravitational attraction is called
    a. the cosmos
    b. a galaxy
    c. a planetary system
    d. a solar system

2. The branch of science that deals with the study of the structure and history of the entire universe is
    a. astronomy
    b. astrology
    c. astrophysics
    d. cosmology

3. The Milky Way, the galaxy within which we live, is
    a. a spiral galaxy
    b. an elliptical galaxy
    c. a quasar galaxy
    d. an irregular dwarf galaxy

4. Large explosive galaxies that pour out vast amounts of energy into space every second are
    a. spiral galaxies
    b. elliptical galaxies
    c. quasar galaxies
    d. irregular dwarf galaxies

5. The statement: "the farther away a galaxy is, the faster it recedes from the Earth" is
    a. Hubble's law
    b. an incorrect statement
    c. one of Newton's important observations
    d. one of Einstein's major observations

6. According to the big bang theory of the origin of the universe
    a. new galaxies are constantly being formed in the space vacated by present-day galaxies
    b. the universe is presently in a steady-state configuration
    c. all galaxies in the universe are moving away from one another
    d. all galaxies in the universe are converging on one another

7. The presence of cosmic microwave background radiation is good evidence for
    a. the big bang theory
    b. the steady-state universe
    c. the account of creation in Genesis
    d. the internal makeup of the Sun

8. Shortly following the big bang start of the universe, the first force to become a separate phenomenon was
   a. the weak force
   b. the strong force
   c. the electromagnetic force
   d. gravity

9. The major occurrence three minutes after the big bang was
   a. the separation of the electroweak and strong forces
   b. the formation of atomic nuclei
   c. the formation of intact atoms
   d. the formation of molecules

10. The abundance of light elements in the universe is good evidence for
    a. the big bang theory
    b. the steady-state universe
    c. the account of creation in Genesis
    d. the internal makeup of the Sun

Fill-In Questions

11. The name of the galaxy that contains the Earth is _____.
12. Edwin Hubble demonstrated that distant galaxies existed by measuring _____ stars.
13. The deviation of distinctive colors emitted by different elements on nearby galaxies toward the long-wavelength end of the spectrum, compared to the light emitted by atoms of the same elements on earth, is called _____.
14. A megaparsec is approximately _____ light-years.
15. The largest groups of galaxies are called _____.
16. A theory of the universe that states that galaxies are constantly being formed in the empty spaces left by receding galaxies is called the _____ universe.
17. The last two forces to separate from one another during the formation of the universe were the weak force and _____.
18. During the elimination of antimatter, antiprotons were annihilated by _____.
19. During the formation of atomic nuclei, elementary particles were formed by the fusion of _____.
20. A large object whose gravitational field is so strong that it does not even allow light to escape is a _____.

Problems

21. Assuming a Hubble constant of 75km/sec/Mpc, what is the approximate velocity of a galaxy 50 Mpc away?
22. If a galaxy is 100 Mpc away, how fast is it receding from us?

Crossword Quiz: Cosmology

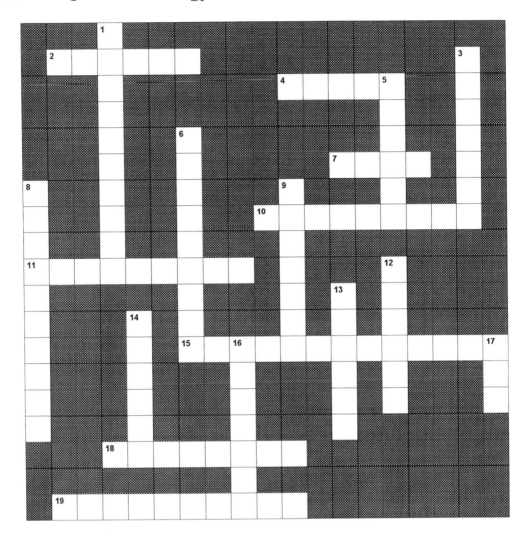

ACROSS

2. A large cluster of stars, dust, and gas
4. Empty spaces between clusters of galaxies
7. Matter in the universe that is undetectable
10. The study of the structure and history of the entire universe
11. The sudden expansion of the universe that happened with the first second of time
15. The largest congregations of galaxies
18. The transposition of distinctive colors emitted by elements in distant galaxies
19. 3.3 million light-years

DOWN

1. A type of galaxy shaped like a football
3. The first force to separate during the formation of the universe
5. The Milky Way is this type of galaxy
6. Dramatic changes in the fabric of the universe during its early history
8. Destroyed within the first second after the big bang
9. The redshift phenomenon is an example of this general process
12. A large galaxy that emits enormous amounts of energy
13. One of the discoverers of the clustered nature of galaxies in the universe
14. Discovered the redshift phenomenon
16. One of the discoverers of the cosmic microwave background radiation
17. The closest star to the Earth

Answers to Review Questions

Multiple-Choice Questions

1. b; 2. d; 3. a; 4. c; 5. a; 6. c; 7. a; 8. d; 9. b; 10. a

Fill-In Questions

11. the Milky Way; 12. Cepheid variable; 13. redshift; 14. 3.3; 15. superclusters; 16. steady-state; 17. electromagnetism; 18. proton; 19. quarks; 20. black hole

Problems

21. 3,750 km/sec; 22. 7,500 km/sec

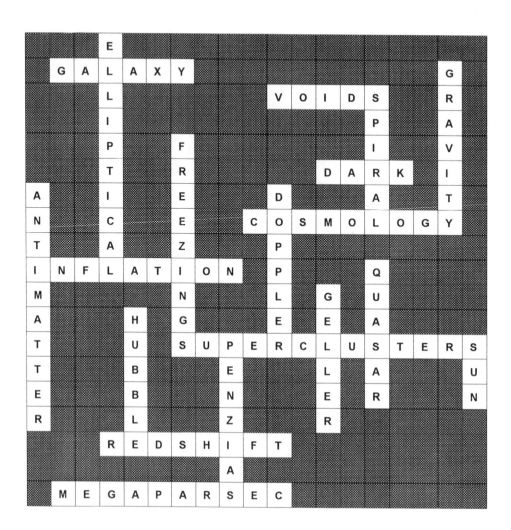

Chapter 16

The Earth and Other Planets

Chapter Review

The solar system includes the Sun, the planets, their moons, and other objects, such as asteroids and comets, that orbit the Sun. All of the planets move in the same direction around the Sun, and in the same rotational plane as the Sun, because they formed about 4.5 billion years from a cloud of dust and gases that was spinning like a frisbee. The Sun contains more than 99% of the mass that was present in the original cloud. At least eight planets condensed through gravitational means in the dust cloud surrounding the sun. The four inner planets, Mercury, Venus, the Earth, and Mars, are known as terrestrial planets because they are relatively small and consist of solid rock materials. The high temperatures of the nearby sun and the strong solar wind swept most of the lighter gases and water vapor from the inner planets into the farther reaches of the solar system, where the gas giants (also called Jovian planets), Jupiter, Saturn, Uranus, and Neptune formed. The outermost planet, Pluto, is small and rocky, and may be an asteroid captured by the Sun's gravitational field. The planets formed by the accretion of dust and gases, depending on which materials were present in the vicinity of the center of formation of each. The heavy metals, probably derived from supernovae (see Chapter 14), tended to concentrate in the interior of the terrestrial planets, while the lighter elements formed an external crust. The Moon, the Earth's large satellite, was probably blasted free from the Earth during a collision with another large body. Unlike the terrestrial planets, the large Jovian planets consist mainly of hydrogen, helium, ammonia, nitrogen, methane, and water. All of the planets, except Mercury and Venus have moons in orbit. In addition to planets, the solar system contains enormous amounts of small, rocky asteroids concentrated in an asteroid belt between Mars and Jupiter. Occasionally, asteroids are captured in the gravitational field of one of the planets, and catastrophic collisions may occur. Comets are icy bodies that are normally found outside of the orbit of Pluto. A comet may pass close enough to the Sun so that the solar wind will deflect enough of its material to produce a luminous tail that may appear spectacular from the earth. Small pieces of debris that fall into the Earth's atmosphere may produce bright meteors ("shooting stars") as they burn up. A meteorite is a rock that reaches the Earth's surface.

Learning Objectives

After studying this chapter, you should be able to:
(Other objectives may also be assigned by your instructor)

1. List the planets of the solar system, and distinguish between the terrestrial planets and the Jovian planets.
2. Discuss the formation of the solar system from a gas-cloud nebula.
3. Describe the formation and early history of the Earth.

4. Discuss the internal structure of the Earth.
5. Describe how a collision with another planetoid could have blasted the Moon from the Earth.
6. Discuss the differences in the atmospheres of the planets.
7. Describe some of the discoveries made by satellites sent to the Jovian planets and their moons.
8. Distinguish between asteroids and comets, and discuss how they might affect the Earth.
9. Describe the differences among meteoroids, meteors, and meteorites.

Key Chapter Concepts

- The solar system includes the Sun, the planets and their moons, and all other objects gravitationally bound to the Sun.
  - All of the planets move in orbits in the same direction around the sun, and this direction is the same as that of the rotation of the Sun.
  - All the orbits of the planets are in more or less the same plane.
  - Virtually all of the mass of the solar system is contained within the Sun.
  - The terrestrial planets include Mercury, Venus, the Earth (and its Moon), and Mars.
  - The Jovian planets include Jupiter, Saturn, Uranus, and Neptune.
  - Pluto, a small, rocky planet, may be an asteroid captured in the Sun's gravitational field.
  - Interspersed with the planets are moons, asteroids, and comets.
- The solar system formed from a large cloud of dust and gas called a nebula.
  - Under the influence of gravity, the nebula slowly started to rotate and collapse upon itself, spinning faster and faster.
  - The rotation created a flattened disk with a central mass of matter. Formation of the Sun at the center of the disk was discussed in Chapter 15.
  - Gases and solid matter accumulated at different places in the disk, resulting in the formation of the planets.
  - The inner planets formed from materials that remained solid at high temperatures.
  - The Jovian planets formed from materials that remained liquid or solid only at the very cold temperatures distant from the Sun.
  - Asteroids and comets remain in the solar system as smaller bodies.
- The Planets evolved in their own distinctive ways.
  - During the great bombardment period of the Earth's formation, a constant rain of solid particles rained down in the earth.
  - The early earth was hot enough to melt nearly completely, and the heavy materials sank under the force of gravity toward the center of the planet. This produced a differentiation of the planet into several layers of material. An inner core (3,400 kilometers thick) made primarily of iron and nickel, is surrounded by a rock mantle almost 3,000 kilometers thick. The crust is relatively thin (10-70 kilometers), and consists of the lightest materials.
  - The Moon appears to have been blasted from the Earth before it hardened by a collision with a large asteroid or smaller planet.
  - The other inner planets formed similar to the Earth.
  - The outer planets accumulated gasses and lighter materials, and although layered, do not have a well-defined solid surface like the Earth and Moon.
- The atmosphere of the Earth has undergone much change during its history.
  - Outgassing, the release of gases from the solid materials of the earth, including volcanic eruptions, contributed greatly to the early atmosphere.
  - It appears that the early atmosphere was composed primarily of methane (CH_4), ammonia

(NH_4), carbon dioxide (CO_2), hydrogen (H_2), and water (H_2O). Much of these gases escaped the gravitational pull of the Earth.
- Much of the water vapor appears to have come from persistent bombardment of the atmosphere by icy comets that continues today.
- Oxygen has been added to the atmosphere mostly by the activities of green plants.
- The outer solar system consists of the Jovian planets, their moons, asteroids, and comets.
  - The Jovian planets consist mainly of liquid or solid materials that are normally gases on the Earth.
  - The moons of these planets are solid, and may even display volcanic activity.
  - Pluto is a small, rocky planet that may be an asteroid that was captured by the gravitational field of the Sun.
  - Asteroids are small, rocky bodies in orbit around the Sun, most in a broad band between Mars and Jupiter.
  - Comets are "dirty snowballs" that orbit the sun outside the orbit of Pluto.
  - Meteoroids are small pieces of space debris in orbit around the Sun. A meteoroid becomes a meteor (shooting star) when it enters the atmosphere and burns up. A meteorite is a meteor that reaches the surface of the Earth.

Key Individuals in Science

- Clyde Tombaugh discovered the planet Pluto on February 18, 1930.
- Pierre Simon Laplace (1749-1827) first proposed the nebular hypothesis for the formation of the solar system.
- Jan Oort (1900-1992) first postulated the existence of a cloud of comets outside of the orbit of Pluto.

Key Concept: The Planetary System

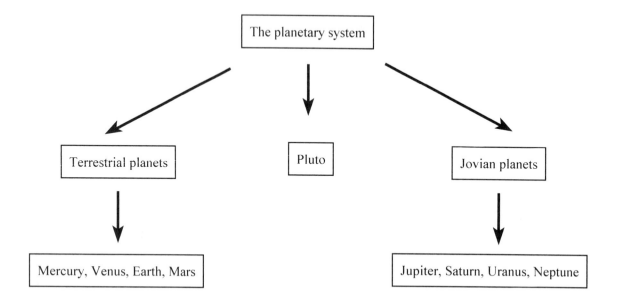

Key Concept: The Nebular Hypothesis of Solar System Formation

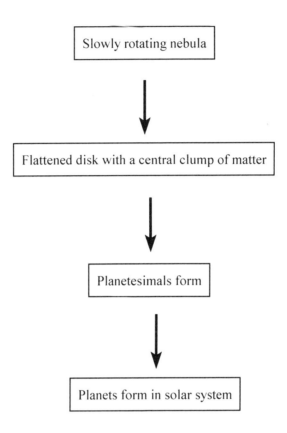

Questions for Review

Multiple-Choice Questions

1. Which of the following is <u>not</u> a terrestrial planet?
   a. Venus
   b. Uranus
   c. Mars
   d. Mercury

2. Which of the following is a terrestrial planet?
   a. Jupiter
   b. Uranus
   c. Neptune
   d. Mars

3. Which of the following is a rocky planet?
   a. Pluto
   b. Jupiter
   c. Neptune
   d. Saturn

4. Clyde Tombaugh was the first individual to identify the planet
   a. Pluto
   b. Jupiter
   c. Neptune
   d. Saturn

5. Which of the following is the largest planet?
   a. Saturn
   b. Uranus
   c. Jupiter
   d. Neptune

6. Another name for a "shooting star" is
   a. meteor
   b. comet
   c. asteroid
   d. planetesimal

7. The inner core of the Earth is formed primarily of
   a. mantle rocks
   b. elements containing oxygen
   c. silicon and magnesium
   d. iron and nickel

8. An icy object orbiting the solar system beyond the orbit of Pluto is
   a. a comet
   b. an asteroid
   c. a meteor
   d. a meteorite

9. The small rocky bodies found in a broad band between Mars and Jupiter are
   a. comets
   b. asteroids
   c. meteors
   d. meteorites

10. A small piece of space debris that reaches the surface of the Earth is
    a. a comet
    b. an asteroid
    c. a meteor
    d. a meteorite

Fill-In Questions

11. Comets orbit the Sun in a region outside the orbit of Pluto called the _____.
12. The largest moon in the solar system is _____, one of Saturn's moons.
13. One of Jupiter's moons, _____, has active volcanoes.
14. The closest to Earth of the gas giants is _____.
15. The release of gases from rocks in the crust of the Earth is called _____.
16. The layer of material under the crust of the Earth, and above the liquid outer core is called the

    _____.
17. The _____ hypothesis is the current scientific explanation of how the solar system formed.
18. A cloud of gas and dust in the Milky Way is called a _____.
19. The planet with the greatest number of moons is _____.
20. The _____ includes the Sun and the planets and their moons.

Crossword Quiz: The Earth and Other Planets

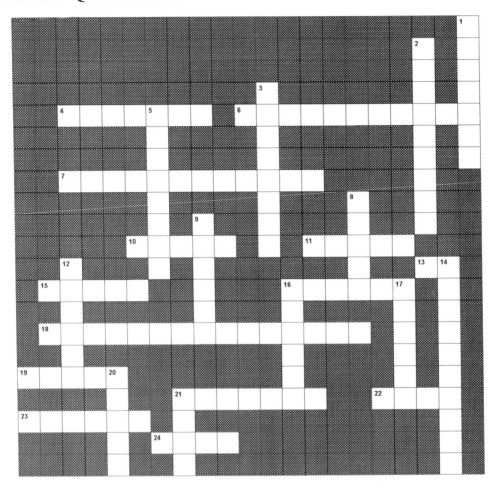

ACROSS

4. One of the gases of the Earth's early atmosphere
6. The period of time during which Earth was pelted with rocks from outer space
7. Small aggregations of rocks found in the nebula forming the solar system
10. This large moon of Saturn may have frigid lakes of methane on its surface
11. The farthest planet from the Sun
13. This Jovian moon has active volcanoes
15. A dirty snowball of outer space
16. The intermediate layer of the Earth
18. The formation of several structural layers during the evolution of the Earth
19. The second planet from the Sun
21. The closest planet to the Sun
22. The planet next farthest from Earth and the Sun
23. A cloud of dust and gases found within the Milky Way
24. The innermost layer of the Earth

DOWN

1. The largest planet in the solar system
2. A piece of planetary debris that reaches the surface of the Earth
3. He discovered the planet Pluto
5. A small, rocky body in orbit between Mars and Jupiter
8. The outer surface of the Earth
9. The planet that is distinguished by its extensive ring system
12. A term referring to the giant gas planets
14. The release of gases from rocks in the crust of the Earth
16. A shooting star
17. This Jovian moon has blocks of frozen ice on its surface
20. Pertaining to the system of planets that surrounds the Sun
21. The satellite of the Earth

Answers to Review Questions

Multiple-Choice Questions

1. b; 2. d; 3. a; 4. a; 5. c; 6. a; 7. d; 8. a; 9. b; 10. d

Fill-In Questions

11. Oort cloud; 12. Titan; 13. Io; 14. Jupiter; 15. outgassing; 16. mantle; 17. nebular; 18. nebula; 19. Saturn; 20; solar system

Chapter 17

Plate Tectonics

Chapter Review

The Earth is a dynamic structure, and it is changing constantly, both internally and externally. Mountains arise due to buckling of the crust, and they are eroded by rain and wind. Two of the most prominent activities that provide evidence of crustal change are volcanoes and earthquakes. Energy transmitted from the Earth's mantle causes thin plates of the crust to move in what is referred to as continental drift. The plates do not "drift" aimlessly, but the movement, called plate tectonics, is directed by the convection currents of heat in the mantle. The study of ridges and deep trenches in the ocean and the patterns of magnetically oriented rocks in the crust have provided convincing evidence of plate tectonics. The radiometric dating of crustal rocks associated with oceanic ridges provided reinforcing evidence for the fact of continental drift. Different boundaries exist between the crustal plates of the Earth. A divergent plate boundary marks the line where two plates are moving away from one another, while a convergent plate boundary marks the line where two plates are pushed together. A transform plate boundary occurs where one plate scrapes past the other, with no new plate material being produced. The geologic history of North America is closely linked to activity along plate boundaries. The earthquake zone of the Pacific coast is associated with the transform plate boundary of the San Andreas Fault and numerous additional faults in the area of Southern California. The volcanoes of the Pacific coast are associated with hot magma forced to the surface near a convergent zone. The Appalachian Mountains of the east coast are associated with a convergent zone that pushed the crust up in this mountain chain. Erosion over millions of years has produced a rounded chain of mountains that differ significantly from the younger Rocky Mountains of the west. Seismology is the study of earthquakes.

Learning Objectives

After studying this chapter, you should be able to:
(Other objectives may also be assigned by your instructor)

1. Describe the forces that cause mountains to be built up and worn away.
2. Discuss the formation and structure of a volcano, and explain how molten lava reaches the surface.
3. Describe the forces that cause earthquakes to occur, and indicate how the Richter scale is used to quantify the magnitude of earthquakes.
4. Describe the phenomenon of continental drift, and indicate the forces responsible for the movement of tectonic plates.
5. List several lines of evidence for continental drift.
6. Discuss the different types of boundaries that occur between tectonic plates.
7. Describe how vulcanism and continental drift have influenced the geologic history of North

America.
8. Discuss the science of seismology and describe how this study increases our knowledge of the internal structure of the Earth.

Key Concepts

- The surface of the Earth is constantly changing.
  - Mountains are formed by volcanic activity and crustal uplift as tectonic plates collide.
  - Mountains are worn away by the eroding action of streams and rivers.
- A volcano is the outlet for molten rock (magma) in the upper mantle to flow to the surface.
  - The material that flows is called lava.
  - Volcanoes are common in three geological situations: divergent plate boundaries, convergent plate boundaries, and hot spots.
  - Volcanic activity along spreading divergent plate boundaries on the ocean floor provides the major source of new crustal rock formation on the Earth.
  - Volcanic activity along convergent plate boundaries with subduction zones, primarily along the "ring of fire" bordering much of the Pacific Ocean.
  - Hot spots provide routes for magma to flow to the surface in isolated areas such as the Hawaiian Islands, Yellowstone Park, and Iceland.
- Earthquakes occur when rock suddenly breaks along a more or less flat surface, called a fault.
  - The energy released travels out from the quake in the form of a seismic (sonic) wave, causing the ground to rise and fall, and shake.
  - There are two principal types of seismic waves. Compressional or longitudinal waves are like sound: the molecules in the rock move back and forth in the same direction as the wave. Shear waves cause the rock molecules to move up and down perpendicular to the direction of the wave motion.
  - Seismology is the study of the seismic waves generated in an earthquake.
  - The Richter scale measures the magnitude of earthquakes on a comparative scale. Each unit on the Richter scale corresponds to a ten-fold increase in the ground movement; such that an earthquake that measures 6 will have 100 times more ground motion than one that measures 4, and so on.
  - In areas where earthquakes are commonplace, buildings are designed to resist the movements of the quake.
- The continents present on the Earth are associated with crustal plates that move in a process often referred to as continental drift.
  - The Earth is broken up into about a dozen large pieces (and some smaller) called tectonic plates.
  - Each plate is a rigid, moving sheet of rock up to 100 km (60 miles) thick, composed of the crust and part of the upper mantle.
  - On time scales of millions of years, the plates shift about on the planet's surface, carrying the continents with them.
  - The energy for movement of the plates comes from heat released through convection in the mantle.
  - Three main lines of evidence support the concept of continental drift:
    - A study of trenches and ridges on the ocean floor revealed that magma was constantly spewing from the mantle, adding to the Earth's crust, and pushing tectonic plates away from one another.

- A study of the orientation of magnetic fields in ancient rocks on the ocean floor (paleomagnetism) revealed that the Earth's magnetic field had reversed directions over 300 times during the past 200 million years. A comparison of complementary bands of these rocks on opposite sides of a fissure showed that the crust had been added from below on a regular basis.
  - Radiometric dating of rocks on both sides of the Mid-Atlantic Ridge confirmed the conclusions reached from studying paleomagnetism in these deposits.
- Three main types of boundaries separate the Earth's tectonic plates.
  - Divergent plate boundaries are lines where two plates are moving away from one another.
    - Such a boundary on the ocean floor allows basalt lava to erupt from the newly created fissure, creating mountain chains and pushing the two plates apart.
    - Rift valleys, such as the Great East African Rift Valley, are produced by divergent boundaries on continents.
  - Convergent plate boundaries are lines where two plates are coming together.
    - A subduction zone is created when one plate sinks beneath another at a convergent boundary.
    - Deep ocean trenches may be formed along subduction zones.
    - High mountains, like the Himalayas, may form along a convergent boundary.
    - On the edges of continents, deep trenches bordered by high mountains may be formed, such as the Cascade Mountains of the northwestern United States and the Andes Mountains of South America.
  - Transform plate boundaries are lines where one plate scrapes past the other, with no new plate material being produced.
    - The San Andreas Fault in California is an example of this type of boundary.
    - Friction associated with the movement along such a boundary causes rocks to break, resulting in frequent earthquakes.
- North America has experienced many of the tectonic phenomena during its geologic history.

Key Individuals in Science

- Charles Richter of The California Institute of Technology devised a scale to quantitatively measure the strength of earthquakes.
- Francis Bacon (1561-1626) first pointed out that the surface of the Earth was in constant flux.
- Alfred Wegener (1880-1930) first proposed the theory of continental drift in 1912.

Key Concept: Activities Along Tectonic Plate Boundaries

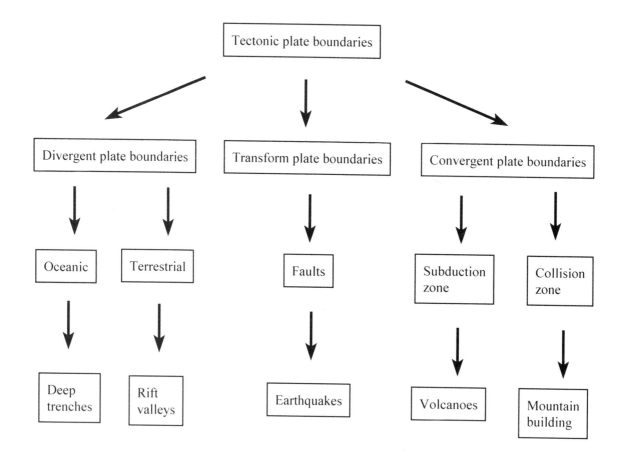

Questions for Review

Multiple-Choice Questions

1. The subsurface molten material in the mantle that enters a volcano is called
   a. lava
   b. magma
   c. fault flow
   d. crustal material

2. The movement that emanates from an earthquake is called a
   a. seismic wave
   b. quake wave
   c. fault wave
   d. magma wave

3. Based on the Richter scale, how much more powerful is an earthquake that measures 10 over an earthquake that measures 9?
   a. two times
   b. ten times
   c. one hundred times
   d. one thousand times

4. The German meteorologist who first proposed the hypothesis of continental drift was
   a. Francis Bacon
   b. Isaac Newton
   c. Charles Richter
   d. Alfred Wegener

5. In the oceans, the crust is composed of
   a. basalt
   b. granite
   c. iron
   d. mud

6. The energy for the movement of tectonic plates comes from
   a. the crust
   b. the mantle
   c. outer space
   d. the Sun

7. A subduction zone is characteristic of
   a. a convergent plate boundary
   b. a divergent plate boundary
   c. a transformed plate boundary
   d. a fault zone

8. The study of earthquake waves is called
   a. geology
   b. seismology
   c. vulcanism
   d. continentalism

9. The study of paleomagnetism and the reversals of magnetic rocks supports
   a. continental drift
   b. geology
   c. seismology
   d. continentalism

10. A hot spot in the Earth's crust is often associated with
    a. a volcano
    b. a fault line
    c. an oceanic trench
    d. an oceanic ridge

Fill-In Questions

11. About 200 million years ago, all of the continents of the Earth were united in a large land mass called _____.

12. In the lower region of the lithosphere, original rocks called _____ may lay encased in the cooled magma underneath a volcano.

13. The subsurface molten rock at the base of a volcano is called _____.

14. Once erupted onto the surface of the Earth, molten rock from a volcano is called _____.

15. An oceanic wave generated by an offshore earthquake is a _____.

16. The theory of continental drift was first proposed by _____.

17. The dense rock that forms the surface of oceanic plates is called _____.

18. About three-quarters of the Earth's surface is covered by _____.

19. The _____ was a vessel that mapped the magnetic rocks on the ocean floor.

20. A _____ is a geologic feature that forms in a transform plate boundary.

Crossword Quiz: Plate Tectonics

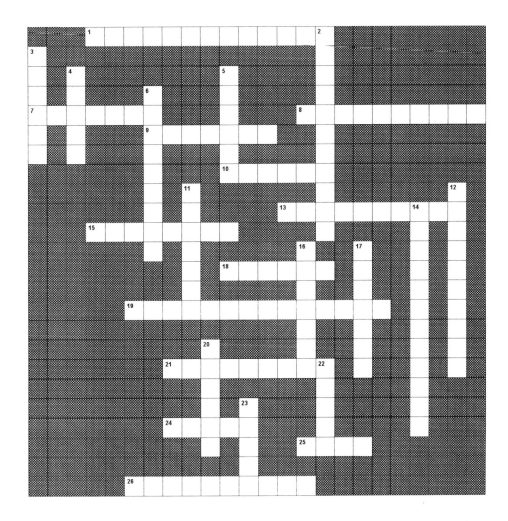

ACROSS

1. The type of seismic wave in which the molecules in the rock move back and forth in the same direction as the wave
7. A large oceanic wave created by an offshore earthquake
8. The process that conveys heat from the mantle to the crust of the Earth
9. A vent in the crust of the earth that allows molten rock to reach the surface
10. A deep fissure in the ocean bottom
13. Movement associated with the release of energy when rocks break in the Earth's crust
15. The interface between two tectonic plates
18. He invented the scale for measuring earthquake intensity
19. The study of the positions of the magnetic poles through time
21. A type of boundary between two tectonic plates that are scraping past one another

24. A seismic wave that generates molecular movement perpendicular to the direction of movement of the wave
25. Molten rock that spews out from a volcano
26. A boundary zone in which one tectonic plate slides underneath another

DOWN

2. Another name for the crust of the earth
3. Blocks of the Earth's crust that move around on the mantle
4. A fissure that develops in a transform plate boundary
5. The crustal material that covers oceanic tectonic plates
6. A plate boundary marked by movement of tectonic plates away from one another
11. Crustal material that forms the covering of terrestrial tectonic plates
12. The study of earthquakes
14. Part of the upper mantle of the Earth
16. He first proposed the theory of continental drift
17. A term used to describe waves of movement in the Earth associated with earthquakes
20. The continental mass that included all of the present continents in a single entity
22. Molten rock under an active volcano
23. A term used to describe the movement of tectonic plates

Answers to Review Questions

Multiple-Choice Questions

1. b; 2. a; 3. b; 4. d; 5. a; 6. b; 7. a; 8. b; 9. a; 10. a

Fill-In Questions

11. Pangea; 12. xenoliths; 13. magma; 14. lava; 15. tsunami; 16. Wegener; 17. basalt; 18. oceans; 19. Pioneer; 20. fault

Chapter 18

Cycles of the Earth

Chapter Review

Except for materials that bombard the earth in the form of meteorites or icy comets, the Earth has a vast, though finite number of atoms. Since no new atoms are being created on the Earth, all of the materials present today have been recycled over the millennia, and will form new objects in the future. The Earth has three major cycles: the atmosphere, the oceans, and the soil. The atmospheric cycle involves the movement of gases and water vapor in weather and climatic cycles. Most of the circulation of the atmosphere is due to the rotation of the planet and the Sun's energy heating the gases of the atmosphere. Storms and weather patterns result from jet streams, prevailing wind currents, and trade winds. The hydrologic cycle describes the evaporation of water from the oceans and other water sources into the atmosphere, and its subsequent return to the ocean and land masses in the form of rain and snow. Large quantities of water may remain locked in glaciers, ice caps, and underground reservoirs for long periods of time. Over thousands of years, glaciers have formed and retreated over much of the land masses close to the poles. The oceans undergo physical mixing when the Earth's rotation and prevailing winds cause currents to flow from place to place. Chemicals cycle between the ocean and the land through rain and sedimentation. Undersea vents and volcanos introduce minerals from the mantle and crust into the ocean when they erupt. The solid portion of the Earth's crust has been formed either through the igneous activities of volcanoes or by sedimentary accumulation caused by wind and water. Sedimentary deposits may be turned into metamorphic rocks if they are buried and subject to intense pressure and heat. All of the cycles described in this chapter are interdependent, and all influence the operation of the others.

Learning Objectives

After studying this chapter, you should be able to:
(Other objectives may also be assigned by your instructor)

1. List the three great cycles of Earth materials.
2. Describe the factors that influence the circulation of the atmosphere.
3. Discuss the differences between weather and climate.
4. List the more common types of storms that move around the planet.
5. Discuss the factors that influence the movement of water from the atmosphere, through the oceans, land masses, and living organisms.
6. Describe Milankovitch cycles and indicate the long-term climatic phenomena they explain.
7. Discuss the factors that determine the patterns of ocean currents.
8. Discuss the recycling of chemicals between the ocean and the land.

9. Differentiate among igneous, sedimentary, and metamorphic rocks.
10. Discuss the interdependence of the Earth cycles.

Key Concepts

- There is no new matter being created on the planet today. Except for the bombardment by meteorites, icy comets, or asteroids, all of the atoms present on the earth are already here.
- The materials present on and in the earth are recycled through several pathways.
- The hydrological cycle describes the movement of water between the oceans, land masses, and the atmosphere.
    - It was once thought that the total amount of water on the Earth's surface has stayed roughly the same since the water formed. However, recent evidence indicates that icy comets are continually bombarding the atmosphere with enough water to account for the total amount of Earth's oceans since the planet cooled sufficiently for water vapor to condense.
    - Water is stored in several major reservoirs.
        - Ice caps at the north and south poles store large quantities of water for long periods of time, where it is not available to actively enter the hydrologic cycle.
        - Glaciers, mainly at the south pole and in Greenland, also store great quantities of water.
        - The oceans and freshwater lakes store water.
        - Underground aquifers store approximately 98% of the fresh water on the Earth.
    - Short-term transfer of water occurs between oceans and the land in the form of rain and snow, and runoff in streams and rivers. Terrestrial life depends on this short-term cycling for water.
    - Ice ages are cyclic periods of formation of large glaciers in temperate regions. These cycles, called Milankovitch cycles, are determined by variations in the Earth's rotation and orbit that cause slight changes in the amount of solar radiation absorbed by the Earth.
    - Ocean currents are caused by the rotation of the earth, and the absorption of heat by the atmosphere and oceans. Ocean currents physically circulate ocean waters.
    - Chemicals are introduced into the ocean as volcanoes and undersea vents erupt, spewing new materials from the mantle and deep crust of the earth.
    - Chemicals continually cycle between the ocean and land.
        - Chemicals may settle out to the ocean bottom, or be incorporated into the bodies of marine organisms that die and settle to the bottom.
        - Uplift and mountain-building forces drain ocean basins and transform undersea beds into surface land.
        - Erosion by wind and rain dissolves the chemicals in terrestrial soils and rocks, and rivers and streams carry the silt back to the ocean.
        - Different elements have different residence times in the oceans, varying from a few thousand years to hundreds of millions of years.
- The atmospheric cycle circulates gases near the Earth's surface.
    - Heating of the atmosphere from the sun tends to create convection cells in the air.
    - Rotation of the Earth divides these cells into three cells in each hemisphere, and "stretches out" the shape of the air circulation pattern in each cell.
    - The winds at the surface tend to blow from east to west near the equator, creating "trade winds."
    - In temperate zones, the winds blow from west to east. The jet stream separates the warmer air mass of the northern hemisphere from the cold polar air.

- Deviations in the jet stream allow mixing of the polar and temperate masses, creating the weather patterns of the northern hemisphere.
- Some of the violent weather patterns include tropical storms, cyclones, hurricanes, typhoons, tornadoes, and monsoons.
- The "El niño" weather cycle in the Pacific basin results from a reversal of the trade winds, and produces drought conditions in normally moist areas and rainy weather in normally dry areas.
- Even though vast areas of the Earth appear to experience long-term stable climate, there is much evidence for rapid climatic change in the past.
- Rocks and soils are formed, destroyed, and altered in a process called the rock cycle.
  - Igneous rocks solidify from liquified magma that spews from fissures in the Earth's crust.
    - Extrusive rocks solidify from volcanic eruptions. Basalt and pumice are both examples of extrusive rocks.
    - Intrusive rocks are igneous rocks that harden underground, before they are exposed to air or water. Granite is the most common intrusive rock in the Earth's crust.
  - Sedimentary rocks are formed by the accumulation of underwater deposits under intense pressure and temperature.
    - Sandstone forms from sand-sized grains of quartz, the most common mineral at the beach.
    - Shales and mudstones form from sediments that are much finer-grained than sand.
    - Limestone forms from the calcium carbonate skeletons of sea organisms.
  - Coral reefs are special kinds of rocks formed by the metabolic activities of living microscopic animals.
  - Metamorphic rocks are formed as sedimentary rocks deep in the Earth's crust are subjected to intense pressure and temperature. Examples of metamorphic rocks include slates, schists, gneisses, and quartzite. Marble is a well-known example of a metamorphic rock.
- The atmospheric, hydrologic, and rock cycles are all interdependent; that is, they all operate continuously, and in concert.
- Although the cycles described in this chapter represent ongoing, long-term modifications to the condition on the Earth, many short-term events also occur, often involving fewer than ten-year periods.

Key Individuals in Science

- Douglas MacAyael, a geophysicist, explained how rocks from Northern Canada were carried out into the North Atlantic ocean by ice sheets.
- Milutin Milankovitch proposed an explanation for the cycles of glacial and interglacial periods that have occurred in temperate regions during the past 2 million years of Earth's history.
- James Hutton (1726-1797) first proposed that some geologic formations seen on the Earth required millions of years to reach their present configuration.

Key Concept: The Atmospheric Cycle

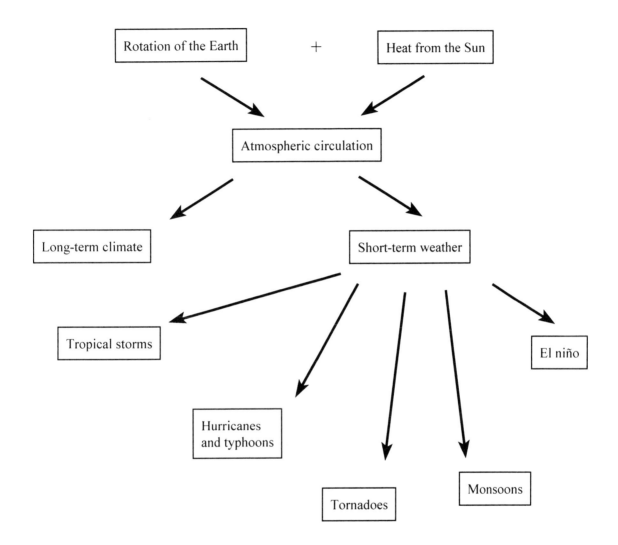

Rotation of the Earth   +   Heat from the Sun

Atmospheric circulation

Long-term climate

Short-term weather

Tropical storms

Hurricanes and typhoons

Tornadoes

Monsoons

El niño

Key Concept: The Hydrological Cycle

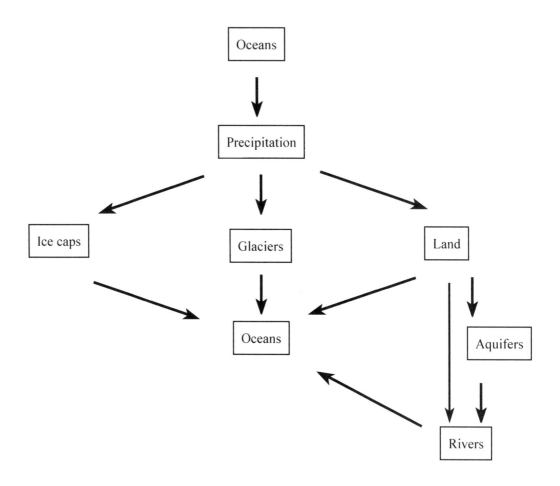

Key Concept: The Rock Cycle

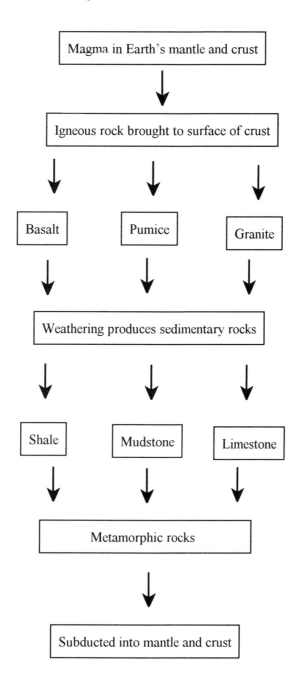

Questions for Review

Multiple-Choice Questions

1. Trade winds south of the equator blow in a _____ direction.
    a. northerly
    b. southerly
    c. easterly
    d. westerly
    e. disorganized

2. Trade winds north of the equator blow in a _____ direction.
    a. northerly
    b. southerly
    c. easterly
    d. westerly
    e. disorganized

3. The jet stream
    a. is a stream of air flowing westerly along the equator
    b. is a stream of air marking the boundary between the northern polar cold air mass and the warmer air of the temperate zone
    c. is a stream of air marking the boundary between the temperate air of the northern hemisphere and the warm humid air above the equator
    d. is a stream of air flowing easterly along the equator
    e. is a stream of air circling the south pole

4. A hurricane that begins in the North Pacific is called a(n)
    a. tropical storm
    b. cyclone
    c. typhoon
    d. tornado
    e. monsoon

5, Approximately 96% of glaciers occur in
    a. Antarctica and Greenland
    b. Canada and Greenland
    c. Russia and Siberia
    d. Canada and the North Polar region
    e. Antarctica and New Zealand

6. Milutin Milankovitch first proposed a theory to explain
    a. the nature of beach erosion
    b. the interdependence of Earth cycles
    c. the age of the earth in billions of years
    d. the role of the jet stream in controlling the weather of the northern hemisphere
    e. the cyclic occurrence of ice ages on the Earth

7. The warm oceanic current that flows northward along the east coast of North America is the
    a. North Equatorial current
    b. Gulf Stream
    c. South Equatorial current
    d. North Atlantic current
    e. Labrador current

8. The average length of time that any given atom will stay in the ocean water before it is removed by some chemical reaction is called the
    a. residence time
    b. dissolution time
    c. mean chemical cycle time
    d. concentration time
    e. precipitation time

9. Basalt is a type of
    a. sedimentary rock
    b. metamorphic rock
    c. intrusive rock
    d. extrusive rock
    e. granite

10. Which of the following is <u>not</u> a sedimentary rock?
    a. shale
    b. pumice
    c. mudstone
    d. limestone
    e. sandstone

Fill-In Questions

11. The _____ cycle is the cycle that describes the movement of water from the oceans to the land and back again.
12. Marble is a type of _____ rock.
13. Rock that is deposited as layers of sand and silt in a lake or ocean is _____ rock.
14. Rocks that solidify under the surface of the crust are _____ rocks.
15. A _____ is a river of moving water within the ocean.
16. The circular motion of the Earth's north pole as it rotates is called _____.
17. An ice age is marked by the advance of _____ across the Earth's surface.
18. An underground storage place for water is an _____.
19. Sudden downdrafts in the atmosphere that create violent air turbulence are called _____.
20. Trends in the weather that are longer than the cycle of seasons are referred to as _____.

Crossword Quiz: Cycles of the Earth

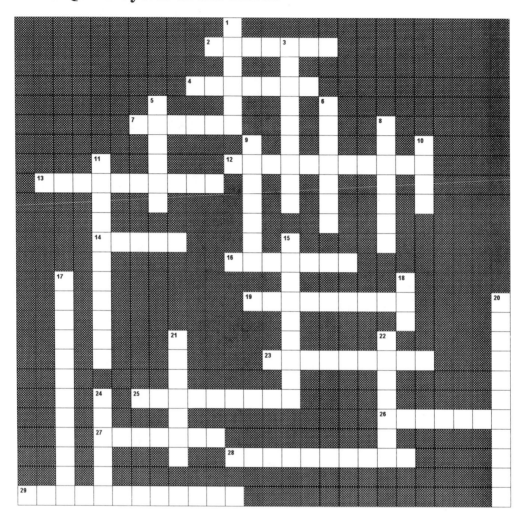

ACROSS

2. A river of water flowing through the ocean
4. Trends in the weather longer than the cycle of seasons
7. A type of metamorphic rock of extraordinary beauty
12. A general type of rock made up of grains of material worn off previous rocks
13. The prevailing winds in the United States
14. A type of sedimentary rock formed from fine-grained materials
16. An intrusive rock formed underneath the surface
19. A tropical storm that forms off the coast of Africa and affects North America
23. A sedimentary rock formed of calcium carbonate
25. The time that an atom remains in the ocean before being removed by a chemical reaction
26. An underground reservoir of water
27. A tropical storm that forms in the North Pacific

28.   The circular wobble that exists in the rotation of the Earth
29.   He proposed the theory explaining the cyclical occurrence of glaciation

DOWN

1.   The igneous rock that forms when magma mixes with a significant amount of water
3.   A general type of rock that solidifies on the Earth's surface
5.   The solidified material that emanates from a volcano
6.   A general name for rocks that solidify from a hot liquid
8.   A large body of ice that slowly flows down a slope or valley under the influence of gravity
9.   The daily changes that occur in rainfall, temperature, and the amount of sunshine
10.   A surface current of water in the ocean that carries cold water from the poles back to the equator
11.   The cycle that explains the movement of air currents over the planet
15.   A general type of rock that solidifies under the crust of the Earth
17.   The cycle that explains the movement of water from the oceans to the land and back again
18.   A stream of air that separates the north polar air from the temperate air of North America
20.   A general name for a type of rock that is formed by intense pressure and temperature deep under the crust of the Earth
21.   Any wind system on a continental scale that seasonally reverses its direction
22.   A rotating funnel of air that descends from storm clouds to the ground
24.   He proposed the ancient time required for geologic processes on the Earth

Answers to Review Questions

Multiple-Choice Questions

1. d; 2. d; 3. b; 4. c; 5. a; 6. c; 7. b; 8. a; 9. d; 10. b

Fill-In Questions

11. hydrological; 12. metamorphic; 13. sedimentary; 14. intrusive; 15. current; 16. precession; 17. glacier; 18. aquifer; 19. wind shear; 20. climate

Chapter 19

Ecology, Ecosystems, and the Environment

Chapter Review

Ecology is the study of the interactions between organisms and their environments. An ecosystem includes the plants, animals, and microorganisms that live in a given area along with their physical surroundings. Ecosystems may vary in size from a balanced aquarium to a lake, meadow, mountain range or continent. Ecosystems consist of both living and nonliving elements. Energy flows through both the living and nonliving parts of an ecosystem. The different levels of energy utilization in an ecosystem are called trophic levels, and the specific specializations within trophic levels work together to form food chains, or food webs. Decomposers act as the limiting trophic level within most ecosystems because they reduce the organic matter in dead organisms' bodies back to inorganic molecules that can be recycled by plants to make new living material. All chemical elements, such as carbon and nitrogen, pass through complex cycles that incorporate both living and nonliving parts of an ecosystem. Every species of organism has a particular mode of life, its ecologic niche. In a stable ecosystem, only one species will occupy a particular ecologic niche at a particular period of time. Stable ecosystems may persist for very long periods of time in a balanced condition of homeostasis. Studies of small ecosystems, indicate that only a fixed number of species exist in a stable ecosystem at a particular time. Environmental changes in either or both of the living and nonliving segments of the ecosystem may bring about disruption and long-term reorganizations. Sometimes, well-intended actions in one part of an ecosystem may have unintended consequences which affect another part of the ecosystem in negative ways. Human activities, such as the generation of trash and its storage in urban landfills, disrupts the natural chemical cycles in the ecosystem. Although very complicated in all its ramifications, recycling of human trash can alleviate many unintended consequences. One very important unintended consequence of human activity is the accumulation of chlorofluorocarbons, a chemical used as a refrigerant, in the atmosphere, which tends to break down the ozone layer. Ozone forms a protective layer that absorbs the harmful ultraviolet rays from the sun before they reach the Earth's surface. Air pollution in the form of nitrogen oxides, sulfur compounds, and hydrocarbons from automobiles and coal-fired industrial plants causes eye-stinging smog in populated areas and acid rain that falls on forests and fields, killing or stunting the growth of plants and animals. Increased production of carbon dioxide by automobile emissions and industrial activities results in a greenhouse effect, an increase in the Earth's temperature due to the accumulated carbon dioxide in the atmosphere trapping heat from sunlight. Scientists and politicians have yet to agree on the long-term effects of global warming on human civilization, or the best methods to alleviate the problem.

Learning Objectives

After studying this chapter, you should be able to:
(Other objectives may also be assigned by your instructor)

1. Define the terms ecology and ecosystem, and give examples of the latter.
2. Define the term ecologic niche, and give an example of one.
3. List the characteristics shared by all ecosystems.
4. Discuss the law of unintended consequences.
5. Discuss the recycling of materials, such as carbon and nitrogen, through both living and nonliving parts of the ecosystem.
6. Define the term trophic level, and relate this term to the concept of a food chain.
7. Discuss the law of unintended consequences.
8. Describe the effect of urban landfills on the cycling of materials through an ecosystem.
9. Discuss the importance of ozone in the atmosphere, and describe the significance of the ozone hole.
10. Describe air pollution, and discuss its relationship to acid rain and global warming.

Key Concepts

- Ecology is the study of the interactions of living organisms with their surroundings.
- An ecosystem includes the plants, animals, and microorganisms that live in a given area together with their physical surroundings.
    - Ecosystems range in size from a balanced aquarium to a coral reef, a forest, or a continent.
    - Ecosystems share several features in common.
        - Ecosystems are varied in the species of organisms they contain.
        - Ecosystems consist of both living and nonliving parts.
        - Energy flows through ecosystems. The source of energy for most ecosystems is the sun, however, some bizarre communities exist deep in the ocean around vents that spew forth heat and chemicals, primarily sulfur compounds.
        - Ecosystems consist of several trophic levels, primarily producers, consumers (both herbivores and carnivores), and decomposers.
        - Only about 10% of the energy available at one trophic level (for example, plants_) is available to the next (the animals that eat the plants). The rest, 90%, is lost as waste heat into space.
        - Matter is recycled through ecosystems continuously. Atoms, such as carbon and nitrogen, pass from the atmosphere, through plants, animals, and decomposers; eventually entering the soil, finally being released back into the atmosphere, where they pass through the same cycles.
        - Every organism occupies an ecological niche, a particular mode of survival. Different organisms continually compete for dominance in their preferred ecological niche.
        - Stable ecosystems achieve a balance among their populations, because there is a finite amount of energy and materials to be shared among the different trophic levels.
        - Ecosystems can be disrupted by changes in the environment or by the extinction of old species and the invasion of new species.

- The law of unintended consequences states that whenever one part of an ecosystem is changed, other parts are affected, often in unpredictable ways.
    - Building levees along the Mississippi River has caused the unintended accumulation of water downstream, with increased flooding.
    - The building of jetties along the shore prevents the natural flow of sand along a beach and results in unintended erosion of beaches down-current from the jetty.
    - The introduction of the Nile perch into Lake Victoria in Africa has wiped out many populations of smaller fish that were important in the diet of local humans, and the necessity to roast the large perch over fires has caused the removal and devastation of forests bordering the lake.
- Human activities have resulted in large-scale assaults on natural ecosystems, with the resulting degradation of the environment.
    - Urban landfills bury large quantities of organic materials under layers of dirt that make it very difficult for decomposers to break down the material and return the chemicals to the natural recycling process.
    - Industrial societies are responsible for the production of large quantities of trash that must be disposed of or recycled.
    - Recycling is a very complex process that requires highly technical solutions to the problems involved in making processed materials available for reuse.
    - The release of chlorofluorocarbons (CFCs) into the atmosphere has reduced the amount of ozone present. Ozone is important as a layer of molecules that absorbs ultraviolet radiation from the Sun, thereby protecting living organisms from exposure to a powerful carcinogenic agent. The situation appears to be especially bad in the southern hemisphere, due to a large ozone hole discovered over the Antarctic.
    - International cooperation will be necessary to reduce the amount of CFCs released into the atmosphere in the future.
    - Air pollution due to automobile emissions and industrial processes, especially the burning of coal, releases significant amounts of pollutants into the air, including nitrogen oxides, sulfur compounds, and hydrocarbons.
    - The presence of pollutants in the atmosphere produces caustic, eye-stinging, smog and when it is washed from the air during an acid rain, the chemicals kill and stunt the growth of plants and animals in the environment. Acid rain also dissolves the outer later of stone buildings, edifices, and works of art.
    - Solutions to the problems of acid rain include reducing the emissions from automobiles and introducing cleaner methods of burning fossil fuels in industrial plants and coal-fired electrical generators.
    - A greenhouse effect occurs on the Earth when carbon dioxide from automobile emissions and industrial activities accumulates in the atmosphere. The excess carbon dioxide absorbs heat from the sun, thereby raising the temperature of the Earth in a process often referred to as global warming.
    - Scientists are not in agreement about the extent of global warming nor its long-term effects.

Key Individuals in Science

- Robert MacArthur and Edward O. Wilson studied the populations on small islands and hypothesized that whenever a new species migrates to an island that already has a thriving and stable ecosystem, it will flourish only if another species becomes extinct.
- Edward O. Wilson and Daniel Simberloff removed all living animals from a group of small islands and studied how they were repopulated from neighboring islands. Once repopulated, the number of species remained about the same as before, but the species of animals were often quite different than before.

Key Formulas and Equations

- The destruction of ozone by chlorine: $2O_3 + Cl + sunlight \longrightarrow 3O_2 + Cl$

Key Concept: Ecology and Ecosystems

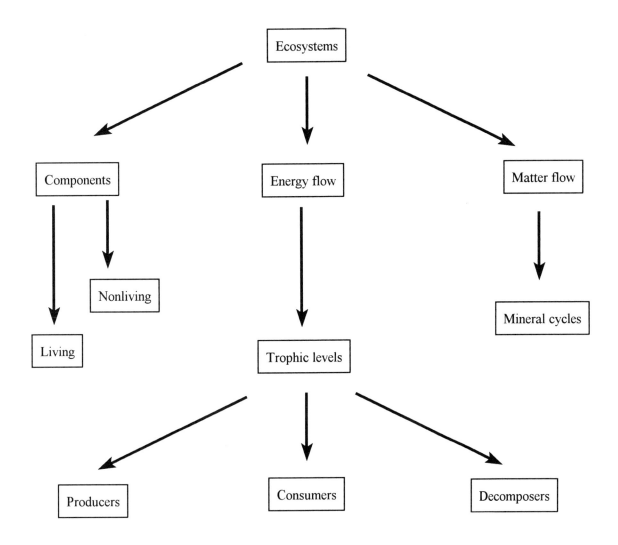

Key Concept: Human Activity Threatens Ecosystems

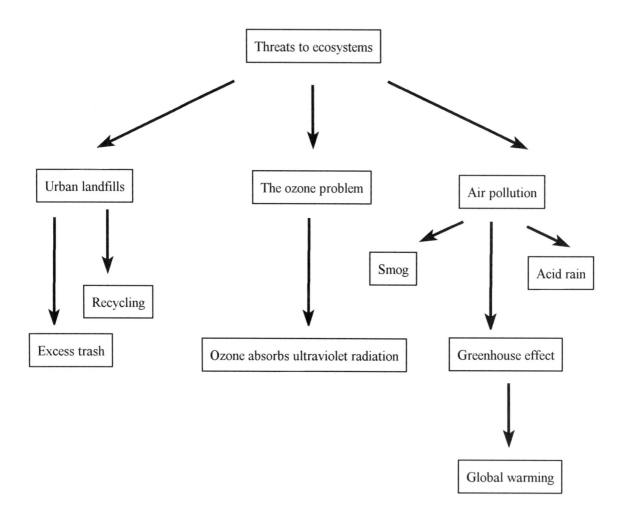

Questions for Review

Multiple-Choice Questions

1. All of the plants, animals, and microorganisms that live in a given area are referred to as a(n)
    a. ecology
    b. community
    c. ecosystem
    d. forest
    e. none of the above

2. The composition of an ecosystem is influenced by
    a. water availability
    b. soil characteristics
    c. average temperature
    d. animal species present
    e. all of the above

3. All of the organisms in an ecosystem form a(n)
    a. ecology
    b. ecological community
    c. species
    d. food chain
    e. trophic level

4. The first trophic level in most ecosystems consists of
    a. photosynthetic plants
    b. herbivores
    c. carnivores
    d. omnivores
    e. decomposers

5. Which of the following trophic levels represents the most limiting factor in an ecosystem?
    a. photosynthetic plants
    b. herbivores
    c. carnivores
    d. omnivores
    e. decomposers

6. In a typical ecosystem, which percentage of energy in a plant is lost to space in the form of waste heat when that plant is eaten by an animal?
    a. 10%
    b. 20%
    c. 50%
    d. 90%
    e. 100%

7. The particular way that an organism obtains matter and energy is called its
   a. ecological niche
   b. ecology
   c. trophic level
   d. environment
   e. ecosystem

8. Ozone in the upper atmosphere is important because
   a. it reduces the temperature of the Earth's surface
   b. it increases the temperature of the Earth's surface
   c. it absorbs ultraviolet radiation from the sun
   d. it prevents the formation of smog
   e. it absorbs chlorofluorocarbons

9. Nitrogen oxides, sulfur compounds, and hydrocarbons combine in the atmosphere to produce
   a. chlorofluorocarbons
   b. air pollution
   c. a special trophic level
   d. a specialized ecosystem
   e. none of the above

10. The buildup of carbon dioxide in the atmosphere appears to be contributing to
    a. acid rain
    b. a decrease in global temperatures
    c. the ozone hole
    d. the greenhouse effect
    e. none of the above

Fill-In Questions

11. _____ is the study of the interactions of living organisms with their physical environment.
12. An _____ includes the plants, animals and microorganisms that live in a given area together with their physical surroundings.
13. All of the living organisms in an ecosystem form an _____.
14. A chart indicating the food relationships among the plants and animals in an ecosystem is called a _____.
15. Trophic levels in an ecosystem indicate the materials and _____ that flow from one species to another.
16. The particular way in which an organism obtains energy and matter is called its _____.
17. The term _____ describes the balance that exists in an ecosystem among their populations.
18. The law of _____ states that it is virtually impossible to change one thing in a complex system without affecting other parts of the system, often in as-yet unpredictable ways.
19. The molecule _____ is made up of three oxygen atoms instead of the usual two.
20. The ozone layer in the atmosphere shields living organisms from _____ from the Sun.

Crossword Quiz: Ecology, Ecosystems, and the Environment

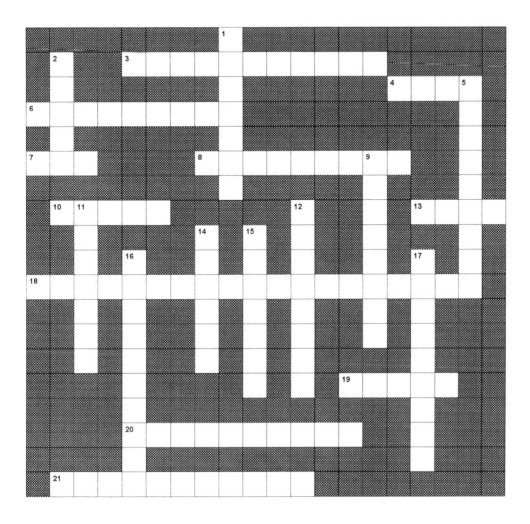

ACROSS

3. This type of radiation is blocked by the ozone layer in the atmosphere
4. This gap in the ozone layer occurs in the southern hemisphere
6. The living organisms in a given area and their physical environment
7. The connections between plants, animals, and microorganisms in an ecosystem
8. A gap in the ozone layer exists above this area
10. This type of fish was introduced in Lake Victoria with unintended consequences
13. This type of rain results from air pollution being washed down onto the land
18. These molecules escape from refrigerators and air conditioners
19. This molecule is composed of three oxygen atoms instead of the usual two
20. The law of _____ consequences
21. This term describes the equilibrium that exists among species in an ecosystem

DOWN

1. This global phenomenon will occur with the buildup of carbon dioxide in the Earth's atmosphere
2. The specific way in which an organism obtains energy and materials
5. Materials that are released from automobile engines
9. This type of radiation is retained on the Earth's surface as carbon dioxide builds up in the atmosphere
11. The study of the interactions between living organisms and their surroundings
12. This element converts ozone molecules to oxygen molecules in the upper atmosphere
14. He is a famous ecologist who studied populations of animals on small islands
15. A level of exchange of matter and energy between organisms in an ecosystem
16. This effect is the result of the buildup of carbon dioxide in the atmosphere
17. The plants, animals, and microorganisms in an ecosystem

Answers to Review Questions

Matching Questions

1. b; 2. e; 3. b; 4. a; 5. e; 6. d; 7. a; 8. c; 9. b; 10. d

Fill-In Questions

11. ecology; 12. ecosystem; 13. ecological community; 14. food web; 15. energy; 16. ecological niche; 17.

homeostasis; 18. unintended consequences; 19. ozone; 20. ultraviolet radiation

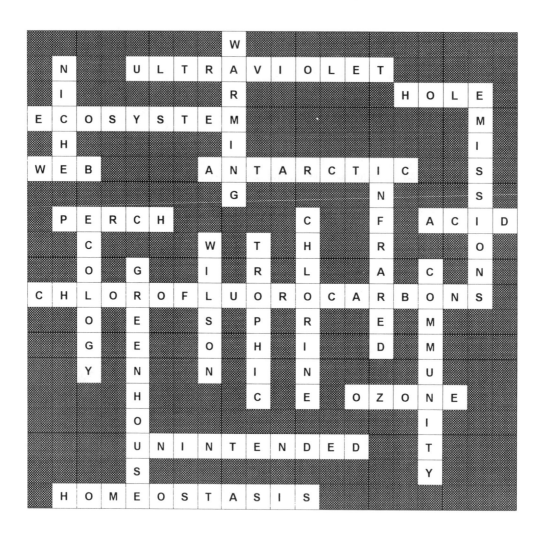

Chapter 20

Strategies of Life

Chapter Review

Biology is the study of living things. Even though living things are the most complex systems studied by scientists, living things operate according to the same laws of nature as everything else we have studied. Different branches of biology study how organisms are adapted to survive in their environments. Even though they are amazingly diverse, all living things are linked by several principles. All living things are part of larger systems of matter and energy. All life depends on chemical reactions that take place in small organized units called cells. Living things regulate their use of energy and respond to their environments. All living things share the same genetic code, which is passed from parent to offspring during reproduction. All living things descended from a common ancestor. All living things may be classified into an orderly framework of units based on shared ancestry as indicated by shared characteristics. Each population of organisms whose members can reproduce fertile offspring are classified as a species, the smallest classification unit. Species are given two names, a binomial, such as *Homo sapiens*, the binomial for the human species. The kingdom is the broadest classification unit. Five kingdoms are currently recognized: Monera, Protista, Fungi, Plants, and Animals. Some scientists currently estimate that there are millions of species present on the earth, most yet unnamed. Different kinds of organisms possess different strategies for survival. Single-celled organisms can absorb their nutrients and excrete wastes directly into the environment, which is often watery. Fungi, such as molds and mushrooms, break down dead organic material and absorb the products into their bodies. Some molds produce byproducts, such as penicillin, that has the capability of killing or limiting the growth of bacteria and other fungi. Plants and algae use the process of photosynthesis to produce food from carbon dioxide and water, using the energy of sunlight. The simplest plants, algae, are either aquatic or grow in moist situations. More advanced plants have mechanisms for picking up water from the environment, usually soil, and distributing it throughout the plant's body. These vascular plants include ferns, conifers, and flowering plants. Animals get their nourishment by consuming molecules produced by other organisms. Among the more conspicuous animals are the invertebrates and vertebrates. Invertebrates include worms, molluscs, crustaceans and insects, while vertebrates are sharks, fish, amphibians, reptiles, birds, and mammals.

Learning Objectives

After studying this chapter, you should be able to:
(Other objectives may also be assigned by your instructor)

1. List the characteristics shared by all living things.
2. Discuss the principles underlying the classification of living things, and list the five major groups of

organisms.

3. Discuss the concept of "binomial nomenclature," and indicate the scientific name for humans.
4. List the main categories that biologists use going from broadest to narrowest to classify living things.
5. Give a rough estimate of the number of species living on the earth today.
6. Discuss the major different strategies for survival used by different types of organisms.
7. Describe different kinds of vascular plants.
8. Distinguish between invertebrate and vertebrate animals.
9. List the variety of vertebrate animals.

Key Concepts

- Biology is the branch of science devoted to the study of living things.
- Different branches of biology study different aspects of the adaptations of organisms for survival.
- In spite of extreme diversity, all organisms utilize the same general survival principles.
    - Living things and their surroundings form complex ecosystems that include the continuous inflow of energy (mainly from the sun) and the recycling of materials between the living and nonliving segments of the ecosystem.
    - Life is based on a series of chemical reactions that occur within small units called cells. Some organisms consist of but a single cell, while others may be composed of trillions of interdependent cells.
    - Living things possess adaptations that allow them to respond to changes in their environment. The adaptations range from as simple as an organism becoming dormant during stressful periods to as complex as an organism shivering to generate heat.
    - All living things share the same genetic code that prescribes the processes used by the organism to survive. The genetic code is inherited from parents during reproduction.
    - All living things produce offspring that are similar in genetic constitution and (usually) appearance to their parents through either asexual or sexual means.
    - All living things descended from a common ancestor. Changes in the genetic information through time and changes in the environment have produced the wide diversity of living things that we now find on the earth.
- Biologists classify living organisms into categories based on their ancestral relationships, and their current similarities and differences.
    - The original Linnaean classification scheme has been modified today to recognize seven levels of relationship among organisms: Kingdom, Phylum, Class, Order, Family, Genus, and Species.
    - A species is a population of similar organisms who can reproduce fertile offspring.
    - Every species is given a distinct name (binomial) consisting of the genus and species name.
    - Scientists estimate that there are several million different species of organisms living on the Earth, most of them yet unnamed.
- Five large kingdoms of organisms are currently recognized.
    - The kingdom Monera contains single-celled organisms that lack an internal structure called the cell nucleus.
    - The kingdom Protista includes many single-celled organisms with a cell nucleus, but also a few multicellular organisms that have a particularly simple body structure.
    - The kingdom Fungi includes multicellular organisms, such as mushrooms and molds that dissolve other dead organisms and absorb the nutrients.
    - The kingdom Plantae includes multicellular organisms that get their energy directly from the

Sun through photosynthesis.
- The kingdom Animalia includes multicellular organisms that get their energy and nutrients by eating other organisms.
- The classification of human beings is as follows:

Kingdom: Animalia
  Phylum: Chordata
    Class: Mammalia
      Order: Primates
        Family: Hominidae
          Genus: *Homo*
            Species: *Homo sapiens*

- Living things display different strategies for survival.
  - Single-celled organisms absorb nutrients and excrete wastes directly from and into the environment, respectively. Some simple multicellular organisms, such as sponges, are aggregations of cells, with each cell essentially functioning on its own.
    - Most single-celled organisms reproduce by one cell splitting into two, however, some have more complex life cycles that involve the union of specialized cells produced at certain times of the year.
  - Fungi, such as mushrooms, molds, and yeast, also absorb nutrients directly from the environment. Many send out slender, thread-like filaments that secrete materials that break down dead organisms.
    - Fungi reproduce asexually by releasing spores that can germinate when they land on a suitable environment. Other fungi have a primitive type of sexual reproduction, which is accomplished by the fusion of two cells in the filaments of different plants.
    - Lichens are combinations of a fungus and an algae that is often capable of growing in very inhospitable habitats. Some lichens can secrete chemicals that break down solid rock, releasing minerals that can be absorbed by the organism.
  - Plants and algae absorb energy from the Sun and store it in organic molecules through the process of photosynthesis (see Chapter 22). This process uses carbon dioxide and water as raw materials and produces oxygen and sugars.
    - Mosses are the simplest type of terrestrial plants. They do not have roots, and absorb water directly through their above-ground structures. They produce food through photosynthesis.
    - Mosses reproduce both sexually (involving sperm and eggs) and asexually (involving spores) in a series of alternating sexual and asexual generations.
    - Vascular plants possess conducting tissues in their roots, stems and leaves that move fluids through the plant.
      - Ferns are the most primitive vascular plants. They produce spores on the undersides of their leaves that can germinate into a small sexual plant. The sexual plant produces eggs and sperm that unite to produce the large, asexual fern plant which is familiar to most people.
      - Gymnosperms, such as pines, cedars, and spruce reproduce through seeds that are borne on cones.
      - Angiosperms are flowering plants. Flowers produce pollen grains, which contain sperm, and ovaries, which contain eggs. Pollination allows fusion of the sperm and egg to produce a seed, which germinates and grows into a new plant.

- Animals are multicellular organisms that must get their nourishment by consuming molecules produced by other life forms. The animal kingdom is very diverse and includes a number of phyla.
    - Invertebrates, organisms without backbones, make up 30 or so phyla. These organisms range in diversity from sponges to earthworms, molluscs, starfish, parasitic worms, jellyfish, crustaceans, insects, and a wide variety of other forms.
    - Vertebrates, organisms that have a spinal cord encased in a backbone, are the most highly advanced animals.
        - The most primitive vertebrates were jawless, eel-like fish, whose living representatives are the lampreys.
        - Sharks and rays are jawed vertebrates whose skeleton is composed completely of cartilage.
        - Bony fish are a class of vertebrates that includes salmon, perch, trout, and most other organisms we identify as "fish." Most breathe using gills, although a few have functional lungs.
        - Amphibians, such as frogs, toads, salamanders, and caecilians are adapted for life in both water and land. The adults have moist skin that requires them to live in areas of high humidity. The adults of most species produce aquatic larvae that develop into adults.
        - Reptiles, such as lizards, crocodiles, turtles, and snakes have tough, scaly skin that prevents the evaporation of water from the body. These forms are fully terrestrial and lay eggs surrounded by a shell that retains water and thus allows development on the land.
        - Birds are the only vertebrates that possess feathers and are adapted for flight. Birds can maintain a high body temperature, and the high metabolic rate necessary for flight.
        - Mammals also are capable of maintaining a high internal temperature. They possess hair, and the females have mammary glands that secrete milk, a fluid that nourishes the offspring until they are capable of feeding themselves. Humans are mammals.

Key Individuals in Science

- Johannes Baptista Van Helmont (1579-1644) performed an experiment with a potted willow tree. The tree weighed 5 pounds when planted, and grew to 169 pounds in five years. He erroneously concluded that all the material used to construct the fabric of the willow tree came from the water he put in the pot.
- Carolus Linnaeus (1707-1778) developed a classification system that grouped all living things according to their shared characteristics.
- Alexander Fleming, a British bacteriologist, discovered the antibiotic penicillin in 1928.
- Howard Florey an Ernst Chain, in 1938, succeeded in producing relatively pure forms of penicillin.
- J. B. S. Haldane (1892-1964), an evolutionary theorist, commented on the relative abundance of beetle species by saying that God "has an inordinate fondness for beetles."
- Howard Morowitz, author of *Entropy and the Magic Flute*, pointed out that Japanese cuisine incorporates more phyla than any other, including such delicacies as seaweeds, sea cucumbers, crustaceans, shellfish, and various bony fish.

Key Concept: Characteristics of Living Things

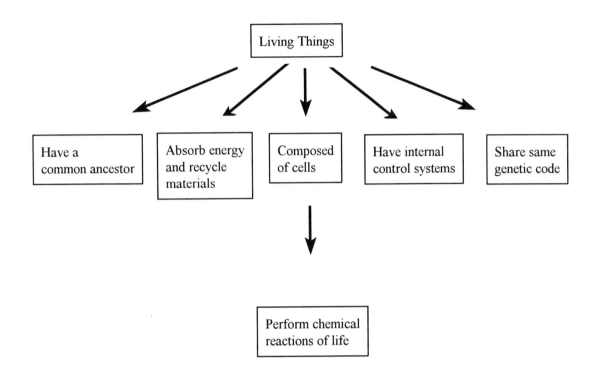

Key Concept: Classification of Living Things

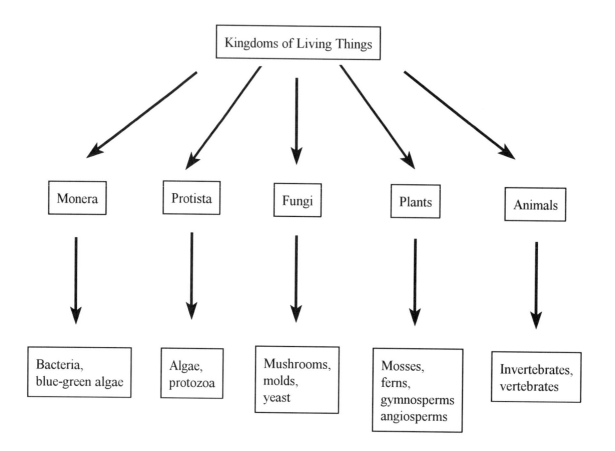

Key Concept: Strategies for Survival

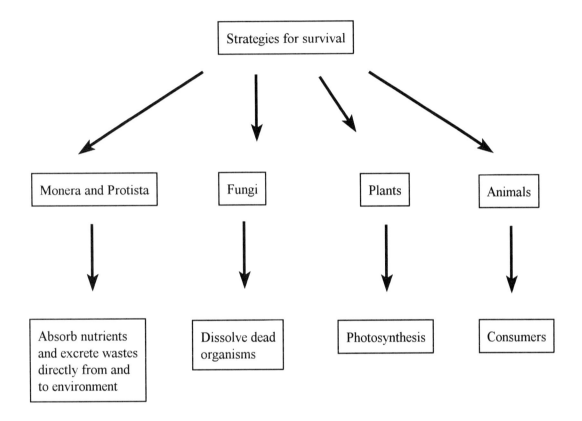

Questions for Review

Multiple-Choice Questions

1. Today, the great majority of biologists study living systems at the level of
    a. molecules
    b. populations
    c. species
    d. kingdoms
    e. organisms

2. In an ecosystem, matter continuously recycles in the system, while _____ flows through it once.
    a. water
    b. oxygen
    c. carbon dioxide
    d. energy
    e. none of the above

3. The smallest units of life are called
    a. molecules
    b. cells
    c. species
    d. organisms
    e. populations

4. After growing a willow tree in a pot for five years, Johannes Van Helmont concluded that the weight gain in the plant came from
    a. the air
    b. the soil
    c. the water
    d. fertilizers applied to the plant
    e. sunlight

5. Carolus Linnaeus is famous because
    a. he pioneered a classification scheme for living things
    b. he discovered penicillin
    c. he purified penicillin after someone else had discovered it
    d. he showed that humans evolved from apes
    e. he performed experiments showing the molecules recycled through ecosystems

6. The broadest classification category of living things is
    a. species
    b. genus
    c. family
    d. phylum
    e. kingdom

7. A population of organisms that can reproduce fertile offspring is a
   a. species
   b. genus
   c. family
   d. phylum
   e. kingdom

8. The kingdom that contains multicellular organisms that get their energy and nutrients by absorbing materials from their environment is
   a. Monera
   b. Protista
   c. Fungi
   d. Plants
   e. Animals

9. To which phylum do humans belong?
   a. nematodes
   b. chordates
   c. vertebrates
   d. invertebrates
   e. primates

10. Which of the following organisms reproduces using "naked seeds?"
    a. mosses
    b. ferns
    c. pine trees
    d. roses
    e. seaweeds

Fill-In Questions

11. The branch of science that gives names to organisms is _____.
12. The specific name for each different type of organism consists of both a _____ and _____ name.
13. The general name for a plant that has specialized conducting tissues in its stems is a _____ plant.
14. The small reproductive structures produced on the undersides of fern leaves are called _____.
15. The process that green plants use to capture sunlight energy is called _____.
16. Penicillin was discovered by _____.
17. The use of two names as the scientific name for an organism is called _____ _____.
18. The kingdom that contains multicellular organisms that get their energy and nutrients by eating other organisms is _____.
19. Humans belong to the Class _____.
20. The scientific name for humans is _____ _____.

Crossword Quiz: Strategies of Life

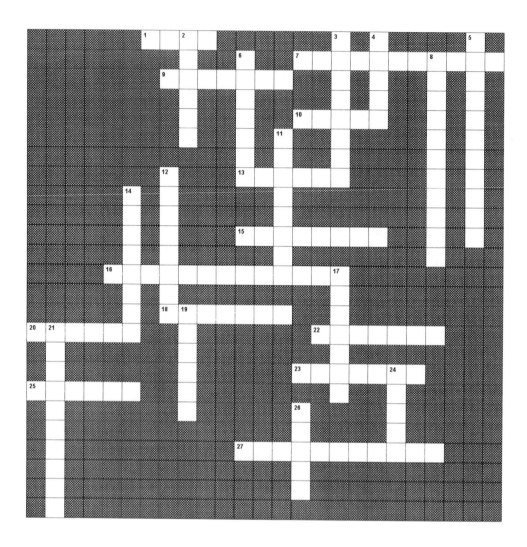

ACROSS

1. The genus name for humans
7. Animals possessing a spinal cord surrounded by a vertebral column
9. Multicellular organisms that obtain energy by consuming other organisms
10. This classification category is composed of several orders
13. Asexual reproductive structures found on the undersides of fern leaves
15. Plants with conducting tissues in their stems
16. Animals that lack a backbone
18. A population of organisms that can reproduce fertile offspring
20. These classification units are grouped together to form an Order
22. The field of science that studies living things

23. The largest classification category
25. This classification category consists of classes
27. This group includes pine and spruce trees

DOWN

2. This group includes bacteria
3. This order of mammals includes monkeys and apes
4. This classification category is grouped into families
5. These animals have a spinal cord surrounded by a backbone
6. These vertebrates have hair
8. This group includes flowering plants
11. This classification group includes green algae and protozoa
12. He developed a system for classifying and naming living things
14. The science of naming organisms
17. The species name for humans
19. Multicellular organisms that utilize photosynthesis to obtain nourishment
21. Insects are included in this group
24. This classification category is composed of families
26. These organisms dissolve dead organisms to obtain nourishment

Answers to Review Questions

Multiple-Choice Questions:

1. a; 2. d; 3. b; 4. c; 5. a; 6. e; 7. a; 8. c; 9. b; 10. c

Fill-In Questions:

11. taxonomy; 12. genus, species; 13. vascular; 14. spores; 15. photosynthesis; 16. Alexander Fleming; 17. binomial nomenclature; 18. animals; 19. mammals; *Homo sapiens*

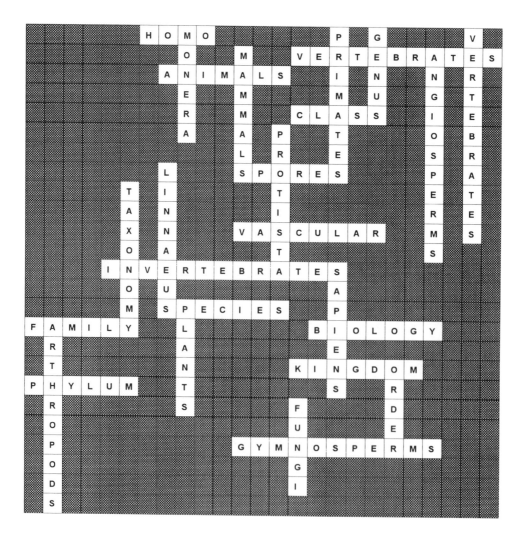

Chapter 21

Molecules of Life

Chapter Review

Organic molecules are characteristic of living things. They are based on the element carbon and the atoms of a few other elements. Just four elements -- hydrogen, oxygen, carbon, and nitrogen -- comprise over 99% of your body's weight. Organic molecules form long chains, branches and rings. Organic molecules are modular and, like building blocks, can be combined to form countless new structures with an extraordinary variety of shapes and sizes. The size and shape of an organic molecule determines the chemical reactions within which it will participate. Four general classes of organic molecules dominate most of life's chemical functions. Proteins, carbohydrates, and lipids are discussed in this chapter. Nucleic acids, such as DNA, are discussed in Chapter 23. Proteins form building materials from which large structures, such as muscles, tendons, hair, and fingernails are formed, and they also serve as enzymes, molecules that control the rate of complex chemical reactions in living things. Proteins are composed of smaller building blocks called amino acids that are linked together by peptide bonds. Proteins may twist and fold in very complex arrangements sometimes involving two or more long chains of amino acids. Carbohydrates are composed of carbon, hydrogen, and oxygen, and range from simple sugars to complex starches. They form many of the solid structures in the body, and are also used as the primary source of energy for cells. Glucose is an example of a simple sugar, a monosaccharide. Sucrose, table sugar, is composed of two monosaccharide units, and is a disaccharide. Polysaccharides are composed of many sugar units joined in long chains. Lipids include fats, oils, waxes, and some specialized substances, such as cholesterol. The carbon atoms in saturated fats are fully bonded to hydrogen atoms, two hydrogens on the sides of each carbon. The carbon atoms in unsaturated fats are not fully bonded to hydrogen atoms, and form double bonds with adjacent carbon atoms. Lipids are used for long-term energy storage. Phospholipids, a form of lipid that is attached to a phosphate molecule, are important in the formation of cell membranes. Even though vitamins and minerals do not supply energy, they are essential to good health because they assist in many chemical reactions. Some elements, such as calcium, are also essential as building materials in the skeleton and other organs.

Learning Objectives

After studying this chapter, you should be able to:
(Other objectives may also be assigned by your instructor)

1. Distinguish between inorganic and organic molecules.
2. Indicate the four most abundant elements in living things.
3. Discuss the relationship between a molecule's shape and its function.
4. Determine the atoms present in a molecule that is written in chemical shorthand.
5. Describe proteins and their relationship to amino acids.

6. Indicate the four levels of protein structure.
7. Discuss the functions of proteins, carbohydrates, and lipids.
8. Describe the three levels of carbohydrate structure: monosaccharides, disaccharides, and polysaccharides.
9. Discuss the structure of lipids, and the difference between saturated and polyunsaturated fats.
10. Indicate the importance of vitamins and minerals in a healthy diet.

Key Concepts

- The molecules found in living things have several characteristics:
    - Most molecules in living systems are based on the chemistry of carbon. Molecules that contain carbon are referred to as organic molecules.
    - Hydrogen, oxygen, carbon, and nitrogen are the four most abundant elements that form the molecules of living things.
    - The molecules of life are modular, composed of simple building blocks linked into a wide variety of shapes and functions.
    - The sizes and shapes of molecules help to determine their behavior, and the chemical roles they fill.
- Chemists use a written shorthand to represent organic molecules. Two practices characterize the diagrams of organic molecules.
    - No hydrogen atoms or bonds to hydrogen atoms are shown in the diagram.
    - Carbon atoms are not shown explicitly.
- Proteins are complex molecules that play important roles in living systems. They consist of carbon, hydrogen, oxygen, nitrogen, and sometimes sulfur and phosphorus.
    - Some proteins form building materials, such as hair, muscles, tendons, and skin.
    - Other proteins function as enzymes, molecules that control the rate of complex chemical reactions in living things.
    - Proteins are constructed of modular building blocks called amino acids.
        - Twenty different kinds of amino acids exist naturally in cells, differing by the specific makeup of a backbone of atoms.
        - Eight of the 20 amino acids are called "essential amino acids" for humans because we cannot synthesize them in our cells. They must be consumed in the diet.
        - At one end of the amino acid is an acidic carboxyl group (COOH). At the other end is a basic amino group (NH_2)
        - Two amino acids can bond together by removing a hydrogen (H) from the amino end of one molecule and a hydroxyl group (OH) from the carboxyl end of the other molecule. This is called a condensation reaction and the bond form between the two amino acids is a peptide bond. A chain of amino acids connected in this manner is called a polypeptide.
    - A polypeptide becomes a protein when it is twisted into an exact three-dimensional shape that allows it to perform its specific function.
        - The primary structure of a protein is its exact sequence of amino acids.
        - The secondary structure of a protein is the first series of spiral twists and folds that it assumes.
        - The tertiary structure of a protein involves further twisting and the formation of cross-links between different parts of the chain.
        - The quaternary structure of a protein results from the linkage of two or more

polypeptides.
- Proteins function as enzymes when they facilitate the bonding together or breaking apart of a chemical bond between other atoms or molecules.
  - The three-dimensional shape of enzymes produces special "sticky spots," called active sites, that attract and hold specific molecules.
  - Once attached to the active site, a chemical reaction occurs that would rarely occur if the reactant molecules could only meet at random.
  - Some drugs function in the body by blocking the active sites of certain enzymes.
- Carbohydrates are made up of carbon, hydrogen, and oxygen. They function both as energy sources and as the building materials for solid structures of living things.
  - The simplest carbohydrates are sugars, such as glucose and fructose.
  - The most common sugars have the chemical formula $C_6H_{12}O$. This unit is called a saccharide unit and simple sugars are called monosaccharides.
  - Disaccharides contain two sugar units combined together, and include such substances as table sugar (sucrose), and milk sugar (lactose).
  - Polysaccharides consist of many sugars linked together in long chains and include such substances as starch and cellulose (plant fiber).
- Lipids include fats, oils, waxes, greases, and other substances that do not dissolve in water.
  - Lipids form the cell membranes that separate living material from its environment.
  - Lipids contain twice as many calories per gram as proteins and carbohydrates, and are used to store energy.
  - Certain lipids have a long, thin backbone with a phosphate group attached to one end of the molecule. Phospholipids are polar molecules, that is to say, one end, the phosphate end, is attracted to water (hydrophilic), while the other end of the molecule is repelled by water (hydrophobic).
  - Saturated fats are solid lipids in which each carbon atom in the chain bonds to two adjacent carbons along the chain and two hydrogens on the sides.
  - Unsaturated fats are liquid oils in which some carbons in the chain will have only three neighbors — two carbon atoms and one hydrogen . This arrangement forms a double bond between the two carbon atoms. A chain with one double bond is monounsaturated, while two or more double bonds yield a polyunsaturated lipid.
  - Cell membranes are formed by a double layer of phospholipids, arranged so that the hydrophobic ends of the molecules point to each other at the interior of the bilayer, while the hydrophilic ends face to the outside.
- Minerals and vitamins are essential chemicals that must form part of our diet.
  - Minerals include all chemical elements in our food other than carbon hydrogen, nitrogen, and oxygen.
    - Among the more important minerals are calcium for strong bones and teeth, and sodium, potassium, chlorine, and magnesium to maintain proper body acidity and control electrical charges in nerve impulses and muscle contractions.
    - Iron is necessary for the formation of hemoglobin in the blood, and iodine is required by the thyroid gland.
  - Vitamins are usually designated by a letter, such as vitamin A, and are needed in minute quantities to participate in chemical reactions in living things.
    - Except for vitamin D, all other vitamins cannot be made in the body and must be taken in with our food.
    - Water-soluble vitamins (those in the C and B categories) must be renewed daily, while

fat-soluble vitamins (A, D, E, and K) can be stored in the liver and other tissues.
- Vitamin deficiencies often result in serious diseases such as scurvy (vitamin C deficiency) and rickets (D deficiency).

Key Individuals in Science
- Friedrich Wohler (1800-1882) was the first scientist to synthesize an organic substance, urea, from inorganic materials.

Key Concept: Protein Synthesis

NH_2 — Amino Acid #1 — COOH   + NH_2 — Amino Acid #2 — COOH

NH_2 — Amino Acid #1 — C0 — NH— Amino Acid #2 — C00H   + $H_2$0

Key Concept: The Structure of Proteins

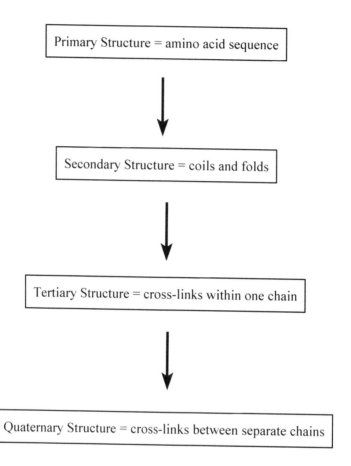

Key Concept: Varieties of Carbohydrates

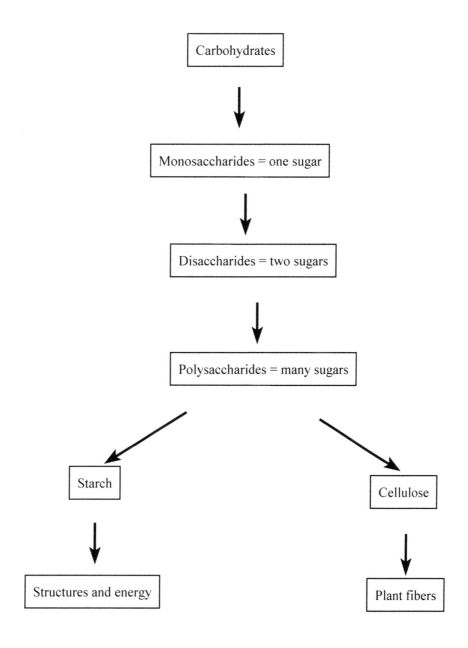

Questions for Review

Multiple-Choice Questions:

1. Organic molecules always contain
   a. water
   b. oxygen
   c. carbon
   d. nitrogen
   e. magnesium

2. In the nineteenth century, the chemist Friedrich Wohler was the first person to synthesize
   a. a monosaccharide
   b. urea
   c. a disaccharide
   d. a protein
   e. an amino acid

3. The building blocks of proteins are
   a. carboxyl groups
   b. amino groups
   c. monosaccharides
   d. disaccharides
   e. amino acids

4. A substance that controls the rate of a chemical reaction in living organisms is
   a. an amino acid
   b. an enzyme
   c. a monosaccharide
   d. an amino group
   e. a carboxyl group

5. A peptide bond is formed between
   a. two amino acids
   b. two carboxyl groups
   c. two amino groups
   d. two monosaccharides
   e. two sugars

6. Cross-links form between different regions of a polypeptide molecule and produce its
   a. primary structure
   b. secondary structure
   c. tertiary structure
   d. quaternary structure
   e. enzymatic structure

7.   The number of amino acids naturally occurring in living organisms is
      a.   10
      b.   20
      c.   30
      d.   40
      e.   50

8.   Carbohydrates consist of
      a.   carbon, hydrogen, and oxygen
      b.   carbon, nitrogen, and hydrogen
      c.   nitrogen, hydrogen, and oxygen
      d.   calcium, carbon, and nitrogen
      e.   phosphorus, carbon, and nitrogen

9.   A lipid whose carbon atoms are attached to two adjacent carbons and two hydrogens on the sides is referred to as
      a.   an unsaturated fat
      b.   a polyunsaturated fat
      c.   a monounsaturated fat
      d.   a saturated fat
      e.   none of the above

10.   Bonds that form between two different polypeptides produce its
      a.   primary structure
      b.   secondary structure
      c.   tertiary structure
      d.   quaternary structure
      e.   enzymatic structure

Fill-In Questions

11.   The first organic chemical to be synthesized in the laboratory was _____.
12.   The bond that forms between two amino acids is a _____ bond.
13.   A straight chain of several amino acids linked together is a _____.
14.   The exact sequence of amino acids that go into a specific protein is called its _____ structure.
15.   An _____ is a molecule that facilitates bonding between two other molecules.
16.   In adults, the eight naturally occurring amino acids that cannot be synthesized in the cell are referred to as _____ amino acids.
17.   Foods that supply amino acids in roughly the same proportions as those in human proteins are called _____ proteins.
18.   Carbohydrates consist of only _____, _____, and _____.
19.   Table sugar, sucrose, is a good example of a _____.
20.   The type of lipid that forms the double layer of the cell membrane is a _____.

Crossword Quiz: Molecules of Life

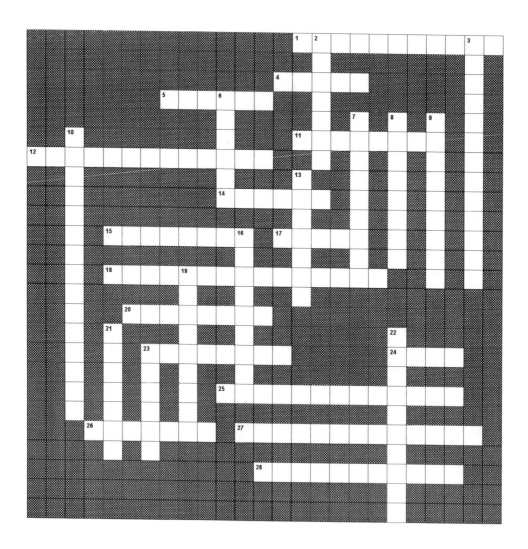

ACROSS

1. A straight chain of amino acids connected together
4. The number of amino acids that humans cannot synthesize
5. This chemist first synthesized an organic molecule
11. Materials in our food, such a sodium, calcium, chlorine, and magnesium
12. The type of lipid that forms the double layer of the cell membrane
14. Sweet substances such as glucose and fructose
15. Organic molecules needed in small quantities designated by letters
17. This group contains a nitrogen atom and is found at one end of an amino acid
18. A carbohydrate containing only one sugar unit

23. This level of protein structure involves bonds between different parts of a single polypeptide
24. The first organic molecule synthesized in the laboratory
25. The process of adding hydrogen to an unsaturated fat
26. The type of condensation bond that forms between two amino acids
27. Organic molecules in living organisms that contain only carbon, hydrogen, and oxygen
28. A type of fat that has double bonds between some of its carbon atoms

DOWN

2. Any molecule that contains carbon
3. A carbohydrate that consists of two sugar modules
6. Fats, oils, and waxes
7. The proper name for plant fiber
8. The COOH group at one end of an amino acid
9. Any amino acid that cannot be synthesized by the human body
10. A carbohydrate composed of numerous sugar molecules
13. The level of protein structure that involves the exact sequence of amino acids in the chain
16. The level of protein structure that involves the twisting and folding of a straight polypeptide chain
19. A fat that does not contain double bonds between its carbon atoms
21. Any molecule that regulates the rate of a chemical reaction
22. The level of protein structure that involves cross-links between two different polypeptide chains
23. The number of amino acids found naturally in living organisms

Answers to Review Questions

Multiple-Choice Questions

1. c; 2. b; 3. b; 4. b; 5. a; 6. b; 7. b; 8. a; 9. d; 10, d

Fill-In Questions

11. urea; 12. peptide; 13. polypeptide; 14. primary; 15. enzyme; 16. essential; 17. high-quality; 18. carbon, hydrogen, and oxygen; 19. disaccharide; 20. phospholipid

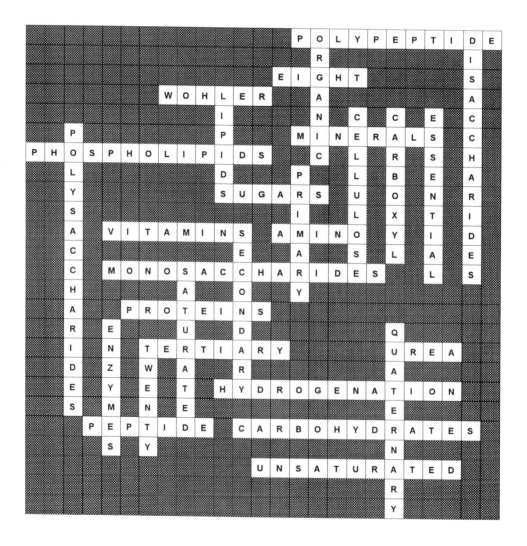

Chapter 22

The Living Cell

Chapter Review

The cell is the smallest unit that is capable of carrying on all of the activities characteristic of living things. Most cells are too small to be seen with the unaided eye, and special instruments, called microscopes, are used to study cells. All cells are surrounded by a cell membrane that consists of a bilayer of phospholipids (see Chapter 20). Embedded in the cell membrane are numerous proteins that act as receptors. These protein receptors bind to specific molecules and mediate the movement of nutrients into the cell and wastes out of the cell. In addition to the soft membrane, plant cells also have a tough cell wall, composed of cellulose and fibers, that adds strength to leaves, stems and roots. The interior of all but the most primitive cells is divided into the nucleus and cytoplasm. The nucleus is surrounded by a double membrane that encloses the genetic material of the cell, its DNA. Within the cytoplasm, cells usually contain other discrete structures, called organelles, such as chloroplasts and mitochondria. Chloroplasts in plant cells use the process of photosynthesis to absorb energy from sunlight and convert it into the chemical energy in the bonds of sugar molecules. Mitochondria absorb molecules derived from glucose, and use oxygen to break those molecules apart and release the chemical energy stored within. The cytoskeleton is a fibrous organelle that moves things around in the cell and gives the cell its shape. All cells carry on the process of metabolism, the exchange of matter and energy with their environment. Once energy is removed from a molecule, such as glucose, it is stored in an energy carrier such as adenosine triphosphate (ATP). Chemical energy is used to attach a terminal phosphate group to adenosine diphosphate (ADP), converting it into ATP. When small amounts of energy are needed in different areas of the cell, the ATP molecule can release its terminal phosphate group, returning to its form as ADP, with the release of energy that can be used to perform a cellular function. Release of energy from a sugar molecule involves several steps. All cells can perform glycolysis, a process that splits the glucose molecule into two pyruvic acid molecules, with the production of six to eight molecules of ATP. Further processing of the pyruvic acids may involve anaerobic reactions that produce alcohol or lactic acid. In cells that can perform aerobic reactions, the pyruvic acids are completely broken down to carbon dioxide and water, releasing enough energy to produce another 30 molecules of ATP. Most cells reproduce by a process called mitosis, in which chromosomes, which have been previously duplicated, are distributed into two daughter cells, each with identical copies of the DNA in the original cell. Meiosis is a specialized type of cell division that produces four cells, each with only half of the original amount of DNA. Meiosis is associated with sexual reproduction, and in more advanced plants and animals, it is involved in the production of sperm and eggs.

Learning Objectives

After studying this chapter, you should be able to:
(Other objectives may also be assigned by your instructor)

1. Discuss the nature of the cell, and indicate who first discovered cells.
2. List the two major kinds of microscopes used to study cells and tell the advantages of each.
3. Describe a typical cell membrane and indicate how it regulates traffic into and out of the cell.
4. Discuss the major organelles of the cell, and indicate their major functions.
5. Describe how plant cells capture energy from the cell and store it in a sugar molecule.
6. Discuss the manner in which cells release energy stored in the sugar molecule and transfer it into the ATP molecule.
7. Distinguish between anaerobic fermentation and aerobic respiration.
8. Describe the process of mitosis, and tell how chromosomes are distributed to daughter cells.
9. Define meiosis and tell how it differs from mitosis.

Key Concepts

- The cell is the smallest unit capable of carrying on all of the activities that we associate with living things.
- In the middle of the nineteenth century, Mattais Schleiden and Theodor Schwann proposed the cell theory:
  - All living things are composed of cells.
  - The cell is the fundamental unit of life.
  - All cells arise from previous cells.
- Cell biology is an important area of biological research today.
  - Cells come in an enormous variety of shapes and sizes, ranging from a few thousandths of a centimeter to the yolk of an ostrich egg. Most cells are about one-hundredth of a millimeter.
  - Cells are studied with magnifying instruments called microscopes.
  - Optical (light) microscopes can magnify objects up to about 1000 times, and resolve details less than a ten-thousandth of a centimeter across.
  - Electron microscopes can magnify objects up to 100,000 times, and their resolving powers can be up to 100,000 times that of an optical microscope.
- Cells are composed of several structures, called organelles.
  - The cell membrane regulates traffic between the inside and outside of the cell.
    - The cell membrane is composed of a bilayer of phospholipids containing proteins and other substances embedded in the membrane.
    - Materials can move across the membrane by diffusion and osmosis.
    - Channels and molecular-sized openings in the membrane allow certain molecules to pass through.
    - Protein receptors embedded in the membrane transport specific molecules through the membrane.
  - Plant cells have a tough cell wall composed of cellulose, surrounding the cell membrane that provides strength to the leaves, stems, and roots.
  - The nucleus is an organelle that contains the genetic material of the cell in the form of DNA.
    - The prokaryotic cells of bacteria and their relatives do not possess a nucleus.

- Eukaryotic cells of single-celled and multi-celled organisms possess a true nucleus.
  - The nucleus is bounded by a double membrane.
- All of the material outside of the nucleus is referred to as the cytoplasm.
- Chloroplasts are organelles found in green plants that perform the process of photosynthesis
- Mitochondria are sausage-shaped organelles where molecules derived from glucose react with oxygen to produce the cell's energy
- The cytoskeleton is a series of channels and fibers that transport materials within the cell and determine the shape of the cell.
- The endoplasmic reticulum is a series of channels that divides the cell into regions, and contributes to protein and lipid synthesis.
- Ribosomes are small spherical bodies that serve as the site of protein synthesis.
- The Golgi apparatus is a specialized part of the endoplasmic reticulum that takes part in the processing of proteins.
- Lysosomes are small spherical bodies that contain digestive enzymes.
- Vacuoles are found in plants, and store water and wastes.
- All cells perform metabolism, the process of exchanging energy and materials with the external environment.
  - Plants use the process of photosynthesis to change sunlight energy into chemical energy and store the energy in the bonds of the glucose molecule.
  - All cells must release energy from the carbohydrate molecules they contain, whether the molecules were manufactured within the cell, as in plants, or consumed, as in animals.
  - Cells store the energy released from the breakdown of glucose, and other organic molecules, into another molecule, called adenosine triphosphate (ATP). ATP serves as the cell's ready supply of energy, providing energy for all every-day chemical reactions.
  - Glycolysis, the first step in energy release in the cell, occurs in the cytoplasm, and breaks the glucose molecule into two pyruvic acids with the release of sufficient energy to produce six to eight molecules of ATP.
  - Fermentation is an anaerobic (without oxygen) process that releases more energy from the pyruvic acids. Prokaryotic cells produce alcohol as a byproduct; while animal cells produce lactic acid as the byproduct of this process.
  - In human cells, the pyruvic acids are processed in the mitochondria in an aerobic (with oxygen) process that releases enough energy to synthesize another 30 to 32 molecules of ATP.
- Cells reproduce themselves on a regular basis using cell division (mitosis).
  - During mitosis, the pairs of chromosomes in a cell (23 pairs in a human cell), previously duplicated, separate and one duplicate of each pair is distributed to a new daughter cell when the cell divides its cytoplasm. The result is two cells, each of which carries a set of chromosomes that are identical to the original.
  - During meiosis, a special type of cell division usually associated with sexual reproduction, the pairs of chromosomes are halved, so that the resulting cells (eggs and sperm) each contain only one-half of each pair. When a sperm combines with an egg in fertilization, the full complement of chromosomes is restored, half from each parent.

Key Individuals in Science

- Robert Hooke (1635-1703), in 1663 first observed the cellulose cell walls of plant cells, and coined the term "cell."
- Anton van Leeuwenhoek (1632-1723) employed superb microscopes of his own design and construction to discover a rich variety of cells, including those in blood, saliva, semen, and the intestines.
- Ernst Ruska, in the 1930s introduced the electron microscope to science.
- Mattais Schleiden and Theodor Schwann proposed the cell theory in the middle of the 19th century.

Key Concept: Cell Structure - The Nucleus

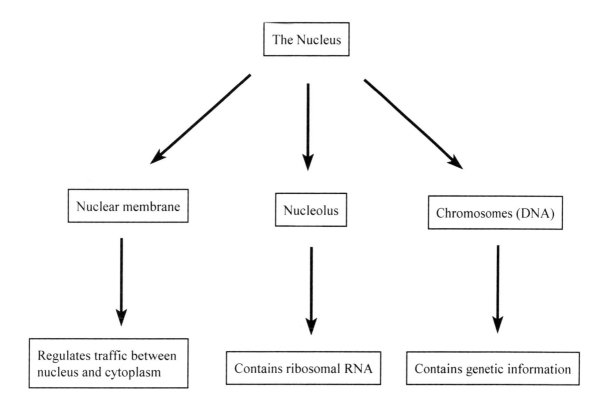

Key Concept: Membranous Cell Organelles

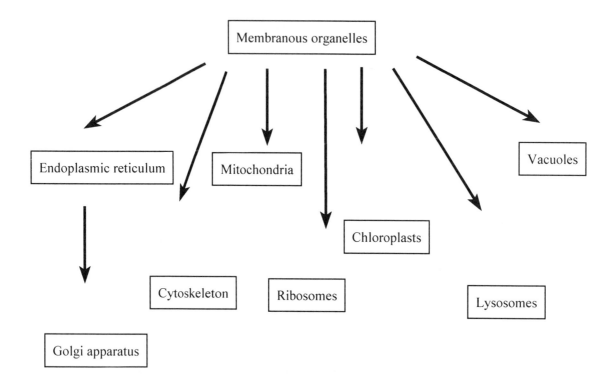

Key Concept: Cell Reproduction

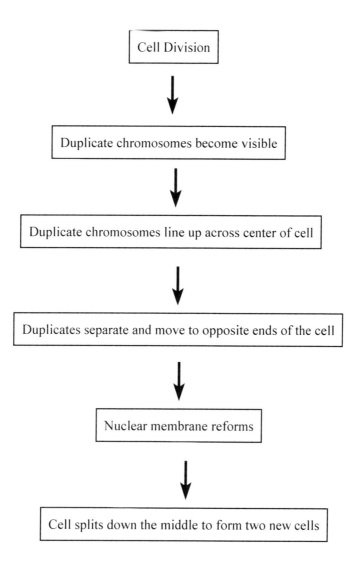

Questions for Review

Multiple-Choice Questions

1. Living cells assume which of the following shapes?
   a. spheres
   b. rectangles
   c. spirals
   d. polygons
   e. all of the above

2. The term cytoplasm refers to
   a. all of the material inside of the cell membrane
   b. all of the material inside of the cell membrane, excluding the nucleus
   c. the fluid that takes up the spaces between the organelles of a cell
   d. the material within the nucleus of a cell
   e. the material surrounding the outside of the cell membrane

3. The major type of molecule in the cell membrane is
   a. phospholipid
   b. proteins
   c. carbohydrates
   d. water
   e. DNA

4. The movement of molecules from a region of higher concentration to a region of lower concentration by ordinary thermal motion is called
   a. diffusion
   b. osmosis
   c. mitochondrial motion
   d. active transport
   e. receptor transport

5. Which type of molecule serves as special receptors in the cell membrane?
   a. phospholipid
   b. carbohydrate
   c. protein
   d. sugar
   e. phosphoprotein

6. The most prominent and important interior organelle of a cell is the
   a. nucleus
   b. mitochondrion
   c. ribosome
   d. chloroplast

7. The organelle that produces energy for metabolism is

a. the nucleus
b. the mitochondrion
c. the ribosome
d. the lysosome
e. the vacuole

8. The site of protein synthesis is
   a. the nucleus
   b. the mitochondrion
   c. the ribosome
   d. the lysosome
   e. the vacuole

9. The organelle in plant cells that converts energy from sunlight into energy-rich sugar molecules is
   a. the nucleus
   b. the chloroplast
   c. the lysosome
   d. the vacuole
   e. the ribosome

10. During photosynthesis, in addition to sugar molecules, another molecule produced is
    a. carbon dioxide
    b. water
    c. carbohydrate
    d. oxygen
    e. pyruvic acid

Fill-In Questions

11. The first step in the extraction of energy from glucose is called _____.
12. In the cytoplasm of the cell, the glucose molecule is broken down into two _____ _____ molecules.
13. Fermentation is the extraction of additional energy from the products of glycolysis in the absence of _____.
14. _____ is the term used to describe the division of one cell into two.
15. DNA is contained within the cell nucleus in structures called _____.
16. Humans have _____ pairs of chromosomes.
17. The special kind of cell division that precedes the production of eggs and sperm is called _____.
18. The _____ of the cell maintains its structure and accomplishes internal transport of material within the cell.
19. Plants cells are surrounded by a _____ _____ in addition to a cell membrane.
20. Prokaryotes are primitive cells that lack a _____.

Crossword Quiz: The Living Cell

Crossword Quiz: The Living Cell

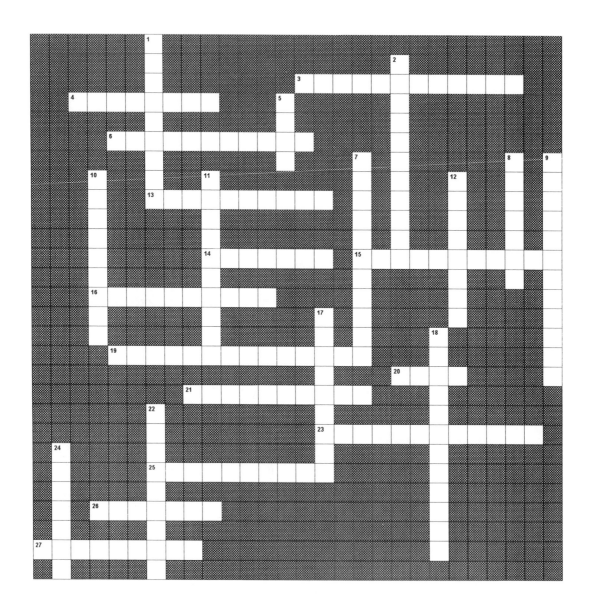

ACROSS

3. This molecule forms the bulk of the cell membrane
4. Waste storage sacks in plant cells
6. The substance that gives plant leaves their green color
13. A device used to observe the interior of cells

19. The mechanism by which plants convert the energy of sunlight into energy stored in carbohydrates
20. The barrier of cellulose that surrounds plant cells
21. The first step in the extraction of energy from glucose
23. The cell organelle that releases energy from the glucose molecule
25. Cells that possess a distinct nucleus
26. The special division that precedes the formation of eggs and sperm
27. The random movement of molecules due to thermal motion

DOWN

1. The fluid material that lies between the cell organelles
2. Thread-like structures composed of DNA
5. The smallest unit of life
7. Cells that lack a distinct nucleus
8. Referring to any process that requires oxygen
9. The process used by yeast cells to produce energy and alcohol
10. These organelles are the site of protein synthesis
11. Distinct structures that lie within the cell
12. The thin material that forms the boundary of a cell
17. Organelles that contain digestive chemicals
18. The organelle that carries on photosynthesis in plants
22. Any process that occurs in the absence of oxygen
24. The process the divides one cell into two cells

Answers to Review Questions

Multiple-Choice Questions

1. e; 2. c; 3. a; 4. a; 5. c; 6. a; 7. b; 8. c; 9. b; 10. d

Fill-In Questions

11. glycolysis; 12. pyruvic acid; 13. oxygen; 14. mitosis; 15. chromosomes; 16. 23; 17. meiosis; 18. cytoskeleton; 19. cell wall; 20. nucleus

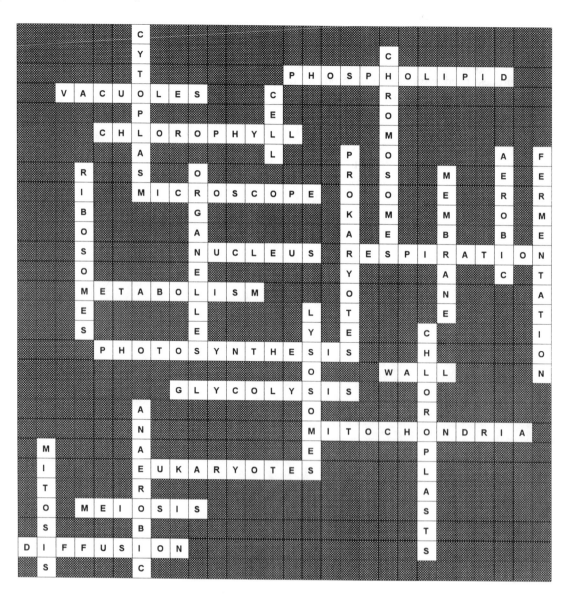

Chapter 23

Classical and Modern Genetics

Chapter Review

Genetics is the study of how biological characteristics are inherited from one generation to the next. Gregor Mendel, an Austrian monk, first revealed the patterns governing the inheritance of physical characteristics in garden peas. He controlled the pollination of flowers in his garden and kept meticulous records of the offspring that resulted from his experiments. He showed that some varieties always bred true to form: tall plants produced tall plants and short plants produced short plants. However, when he experimentally cross-pollinated the tall and short plants, all of the offspring were tall; the factor that produced short plants seemed to have disappeared. Then, when he cross-pollinated the hybrid plants produced from the first experiment, approximately one-fourth of the offspring were just as short as the plants used in the original cross. Based on his experiments, Mendel proposed the existence of a unit of inheritance (we now call it the gene), and that certain traits (such as tallness) were dominant, that is, they were expressed completely when in combination with others (such as shortness) he called recessive. He then proceeded to study the combination of two traits, such as flower color and seed coat characteristics. Mendel's findings led to the establishment of three rules of classical genetics: 1) Physical characteristics or traits are passed from parents to offspring by some, at that time unknown, mechanism (we call it a gene). 2) Each offspring has two genes for each trait — one gene from each parent. 3) Some genes are dominant and some are recessive. When present together, the trait of a dominant gene well be expressed in preference to the trait of a recessive gene. Mendel's findings also led to a quantification of our concepts of inheritance and a realization of how we can use the laws of Mendelian genetics to trace cases of hereditary or genetic disease. We now know that chromosomes are composed of DNA, deoxyribonucleic acid. Another nucleic acid, RNA, ribonucleic acid, also occurs in the nucleus, but not as a structural part of the chromosome. Nucleic acids are polymers composed of long chains of monomers called nucleotides. Each nucleotide consists of a sugar (either ribose or deoxyribose), a phosphate group, and a base. Four bases occur in DNA: adenine (A), guanine (G), cytosine (C), and thymine (T). The base uracil (U) substitutes for thymine in RNA. DNA consists of two strands of nucleotides joined together to form a complete "ladder". Adenine can only form bonds with thymine, and guanine can only form bonds with cytosine. RNA is built in a manner similar to DNA, however, it is only a single strand of nucleotides. Several kinds of RNA operate within the cell simultaneously. Before cell division, DNA replicates by separating the paired bases at the center of the ladder. Exposed bases on each strand can attract and bond with loose appropriate nucleotides in the surrounding fluid to reestablish the double strand. Each reconstituted strand is distributed to a new daughter cell formed during mitosis. The genetic information carried on DNA is transcribed by synthesizing new RNA molecules on open segments of the DNA molecule. Each stretch of DNA carries instructions for only one protein. The RNA is moved out of the nucleus into the cytoplasm. Three kinds of RNA may be formed in this manner: messenger RNA (mRNA), transfer RNA (tRNA), and ribosomal RNA (rRNA). Messenger RNA carries the genetic information for the sequence of amino acids. Transfer RNA reads the message on mRNA and arranges amino acids in the proper sequence for a specific polypeptide. Ribosomal RNA forms the ribosome, the site of polypeptide formation. All living things utilize the same code and the same genetic mechanism to produce proteins. Any change in the normal genetic information is called a mutation, and is usually harmful. Viruses are short lengths of DNA or RNA wrapped in a protein coating , that take over the metabolic activity of a cell, using it to produce more virus particles. The Human Genome Project is a massive scientific attempt to document the exact code of all of the genes located on human chromosomes.

Learning Objectives

After studying this chapter, you should be able to:
(Other objectives may also be assigned by your instructor)

1. Distinguish between purebred and hybrid individuals.
2. Describe the major types of genetic experiments performed by Mendel.
3. Define the term gene.
4. Distinguish between dominant and recessive genes.
5. List the rules of classical genetics.
6. Describe the structure of DNA and RNA and list the three structural differences between them
7. Describe how RNA transcribes and translates the hereditary information contained in chromosomes into proteins.
8. Discuss the genetic code, and indicate how mRNA codons specify certain amino acids.
9. Define the term mutation.
10. Discuss the Human Genome Project.

Key Concepts

- Genetics is the study of how biological information is passed from one generation to another.
- Classical genetics is the quantitative description of hereditary patterns.
    - Organisms that breed true for a certain trait are said to be purebred for that trait.
    - Hybrids are the result of crossbreeding two purebred organisms.
    - Three rules summarize classical genetics:
        - Physical characteristics or traits are passed from parents to offspring by some unknown mechanism — we call it a gene.
        - Each offspring has two genes for each trait — one gene from each parent.
        - Some genes are dominant and some are recessive; the trait of a dominant gene will be expressed in preference to the trait of a recessive gene.
    - Classical genetics allowed the quantitative expression of how traits are inherited.
    - Classical genetics allowed scientists to trace cases of hereditary or genetic disease, such as hemophilia.
- Molecular genetics is the study of the actual hereditary material, nucleic acids. There are two nucleic acids: DNA (deoxyribonucleic acid) and RNA (ribonucleic acid).
- Each nucleic acids consists of a long strand of smaller building blocks, called nucleotides.
    - In DNA, each nucleotide consists of a sugar (deoxyribose), a phosphate group, and a base. There are four nucleotides, determined by the specific base that is present: adenine (A), thymine (T), guanine (G), and cytosine (C).
    - In RNA, each nucleotide consists of the sugar ribose, a phosphate group, and the base uracil

(U) is substituted for thymine.
- The DNA molecule consists of a double strand of nucleotides held together by hydrogen bonds.
  - This double strand resembles a ladder in which the rungs are formed by paired bases.
  - Because of their geometry, the bases are always paired in the following manner: A with T, and G with C.
  - Twisting of the ladder produces a spiral-shaped structure referred to as a double helix.
- The RNA molecule consists of a single strand of nucleotides held together by covalent bonds.
- DNA is replicated prior to every cell division.
  - The first step in replication involves breaking the hydrogen bonds holding the bases together to form the rungs of the ladder, exposing the bases on the two split arms.
  - Exposed bases attract loose nucleotides such that exposed adenine (A) bases will bond with free available thymine (T) based nucleotides, and guanine (G) bases with free available cytosine (C) based nucleotides.
  - Two strands of DNA result, each carrying the identical sequence of nucleotides as the original molecule.
  - A mistake in the replication process, called a mutation, may rarely occur.
  - During mitosis, one copy of each molecule is distributed to the new daughter cells.
  - During meiosis, each daughter cell receives only one chromosome from each pair of chromosomes in the original cell.
  - At fertilization, a full set of chromosomes is restored, but one chromosome in each pair comes from the father's sperm, while the other comes from the mother's egg.
- DNA carries all our genetic information in the nucleus, and in mitochondria and in plants, chloroplasts. Each nuclear gene carries information to produce one specific protein.
  - RNA is used to transcribe (copy) the genetic message on the nuclear DNA and transport the message out to the cytoplasm, where it can be translated into proteins.
  - The process begins when a section of DNA carrying the message unzips, exposing the bases on the nucleotides.
  - RNA nucleotides are attracted to the DNA bases, U to A, C to G, and so forth. The result is a single-stranded RNA molecule with information (like the negative of a photograph) taken from the DNA.
  - Three kinds of RNA are produced.
    - Messenger RNA (mRNA) carries the message indicating the actual identity and sequence of amino acids that will compose the protein. The message is carried in units of three bases, each called a codon. All living things utilize the same suite of codons to specify the same amino acids.
    - Ribosomal RNA (rRNA) leaves the nucleus and forms the ribosome, the actual site of protein synthesis, in the cytoplasm.
    - Transfer RNA (tRNA) transports the amino acids to the ribosome where they are linked together in the sequence specified by the mRNA.
    - Sixty-one of the 64 different tRNAs attach, one each, to the 20 different amino acids needed to produce a polypeptide. The remaining three act as "stop" signs, which end the construction of the polypeptide.
  - Protein synthesis involves all three RNAs.
    - The mRNA and rRNA molecules combine to form the protein factory.
    - Transfer RNA molecules are attracted in turn to individual codons on the mRNA molecule as the ribosome starts at one of the mRNA and moves along its length.
    - As the tRNA molecules are aligned in sequence, the amino acids they are carrying are

linked in the correct sequence that produces that particular protein.
- When the ribosome reaches a stop codon, synthesis ends.
- The polypeptide produced can now be processed by other cell organelles (see Chapter 22) to produce its exact three-dimensional shape.
- Our understanding is rudimentary when it comes to explaining why only certain genes are transcribed at certain times and not others.
- Viruses are short lengths of DNA or RNA wrapped in a protein coating.
  - Viruses cannot survive outside of living cells.
  - Viruses are taken into a cell, where they take over the metabolic machinery of the cell. The virus causes the cell to transcribe its genetic information and produce more viruses.
  - The HIV virus that causes AIDS in humans is particularly insidious in that it subverts and destroys certain cells in the immune system,
- The Human Genome Project is a massive effort to document the exact sequence of all of the DNA in the 23 pairs of human chromosomes.
  - The project uses a technique, called DNA mapping, to discover the nature of the base pairs for each gene.
  - Identification of the location of all of the genes will provide the most fundamental basis for understanding human development and health.

Key Individuals in Science

- An Austrian monk, Gregor Mendel (1822-1884), was the first person to quantitatively describe the process of genetic inheritance. Even though he had no concrete knowledge of "genes," he used experimental methods to show the existence of particulate inheritance and the concepts of dominance and recessiveness.

Key Concept: The Replication of DNA

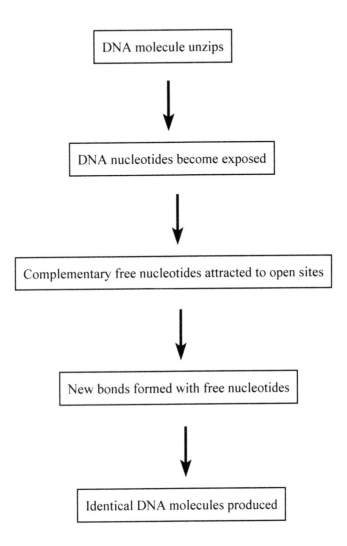

Key Concept: Protein Synthesis

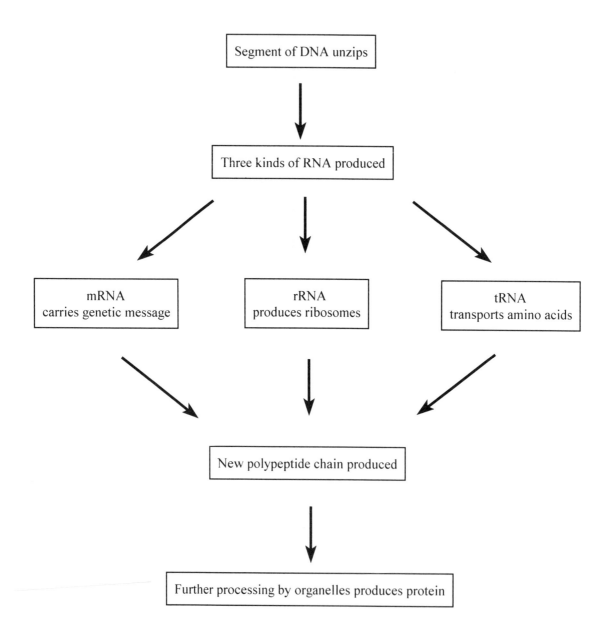

Segment of DNA unzips

↓

Three kinds of RNA produced

mRNA
carries genetic message

rRNA
produces ribosomes

tRNA
transports amino acids

New polypeptide chain produced

↓

Further processing by organelles produces protein

Questions for Review

Multiple-Choice Questions

1. Mendel performed his classical genetics experiments using
    a.   fruit flies
    b.   garden peas
    c.   laboratory mice
    d.   guinea pigs
    e.   human subjects

2. The unit of inheritance is
    a.   viral DNA
    b.   hybrid DNA
    c.   the gene
    d.   messenger RNA
    e.   transfer RNA

3. A genetic characteristic that always appears or is expressed is called
    a.   dominant
    b.   recessive
    c.   hybrid
    d.   purebred
    e.   there are no such traits

4. How many genes for each trait does an offspring receive from each parent?
    a.   one
    b.   two
    c.   for
    d.   eight
    e.   it varies for each trait

5. DNA and RNA are both examples of
    a.   proteins
    b.   lipids
    c.   complex carbohydrates
    d.   nucleic acids
    e.   nucleotides

6. The building blocks of nucleic acids are
    a.   proteins
    b.   lipids
    c.   complex carbohydrates
    d.   nucleic acids
    e.   nucleotides

7. If the sequence of nucleotides on one side of a DNA molecule is A-T-T-G-A-C, the complementary sequence on the other side would be
    a.   A-T-T-G-A-C
    b.   C-A-G-T-T-A
    c.   T-A-A-C-T-G
    d.   A-U-U-C-U-G
    e.   U-A-A-C-U-G

8. If the active sequence on a DNA molecule to be transcribed is C-C-G-A-T-A, the complementary sequence on the transcribed mRNA molecule would be
    a.   C-C-G-A-T-A-
    b.   G-G-C-U-A-U
    c.   G-G-C-T-A-T
    d.   C-C-G-U-A-U
    e.   C-C-G-T-A-T

9. Amino acids are carried in the cytoplasm by
    a.   DNA
    b.   mRNA
    c.   tRNA
    d.   rRNA
    e.   genetic codons

10. The copying of the DNA code onto an mRNA molecule is referred to as
    a.   translation
    b.   transcription
    c.   replication
    d.   mitosis
    e.   meiosis

Fill-In Questions

11. Proteins are actually synthesized at the _____.
12. Any change in the DNA is called a _____.
13. A _____ consists of a short length of RNA or DNA wrapped in a protein coating.
14. The three-nucleotide segments of mRNA that designate specific amino acids are called _____.
15. The nucleotides in complementary strands of a DNA molecule are held together by

    _____.
16. An organism that always produces offspring having the same form of a certain trait is said to be

    _____.
17. A trait that is not expressed when in combination with another form of the same trait is said to be

    _____.
18. In the DNA molecule, cytosine always combines with _____.
19. In the RNA molecule, thymine is replaced by _____.

20. The sum of all information contained in the DNA for any living organism is known as that

organism's _____.

Genetics Problems

21. In squash, a gene for white color (W) is dominant over its allele for yellow color (w). Give the distribution of traits in the following cross:

    Ww x Ww

22. In human beings, brown eyes are usually dominant over blue eyes. Suppose a blue-eyed man marries a brown-eyed woman whose father was blue eyed. What proportion of their children would you predict will have blue eyes?

23. In sheep, white (B) is dominant to black (b). Give the distribution of traits in the offspring resulting from the cross of a pure-breeding white ram with a pure-black ewe.

24. In radishes, the shape may be long (LL), or round (ll). The color of radishes may be red (RR) or white (rr). Give the distribution of traits in a cross of a pure-breeding long red radish with a pure-breeding round white radish.

25. Suppose you crossed two of the radishes that resulted from the cross in #24. What would be the distribution of traits in their offspring?

Crossword Quiz: Classical and Modern Genetics

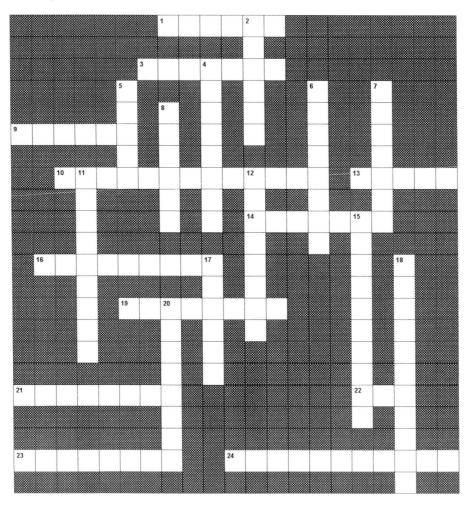

ACROSS

1. The total genetic information contained within any living organism
3. This DNA base bonds to adenine
9. The result of a cross between two purebred individuals
10. The copying of the DNA code onto RNA
13. A short length of DNA or RNA wrapped in a protein coating
14. This base bonds to thymine
16. This type of RNA transcribes the DNA code for protein synthesis
19. An organism that always breeds true for a certain trait
21. The scientific study of the inheritance of biological characteristics
22. The material in the chromosomes that carries the genetic code
23. This base bonds to guanine
24. This sugar is found in DNA

2. This man discovered the pattern of genetic inheritance
4. Any change in the genetic code
5. A unit of the genetic code consisting of three nucleotides
6. This type of trait is always expressed
7. This base bonds to cytosine
8. This RNA base bonds to adenine
11. This type of character is always hidden by a dominant trait
12. This type of RNA carries amino acids
15. Nucleic acids are composed of this monomer
17. This sugar is found in RNA
18. The copying of the DNA molecule prior to mitosis
20. Protein synthesis occurs at this cell organelle

Answers to Review Questions

Multiple-Choice Questions

1. b; 2. c; 3. a; 4. a; 5. d; 6. e; 7. c; 8. b; 9. c; 10. b

Fill-In Questions

11. ribosome; 12. mutation; 13. virus; 14. codons; 15. hydrogen bonds; 16. purebred; 17. recessive; 18. guanine; 19. uracil; 20. genome

Genetics Problems

21. 75% white, 25% yellow; 22. 50%; 23. 100% white; 24. 100% long-red; 25. 9 long-red : 3 long-white : 3 round-red : 3 round-white

Chapter 24

The New Science of Life

Chapter Review

Genetic engineering involves procedures by which foreign genes are inserted into an organism, or existing genes are altered to modify the function of an organism. The procedure employs certain proteins, called restriction enzymes, which cut the DNA molecule in such way that another DNA molecule with different genes that has been cut in a similar fashion can be spliced onto the original molecule. This allows the original cell, and its offspring, to carry on activities associated with the new gene, such as bacterial cells producing human insulin. These techniques have also been applied to producing new strains of crops and agricultural animals with unique characteristics. The analysis of codon sequences in human DNA has allowed law enforcement agencies to increase the accuracy with which perpetrators of crimes are identified. Scientists are now using the techniques of gene therapy to replace defective genes with healthy ones in individuals who are good candidates for this kind of treatment. Viruses appear to be good candidates for carrying the new genes into target cells, because that is their normal mode of operation. Experimental evidence indicates that tumors and cancers are caused by a defective gene (p53) that is involved in determining when a cell will divide. Cancer cells with a defective p53 gene cannot control their rate of division. Superoxide ions are particularly effective in producing genetic damage. DNA has a repair system to correct mistakes that occur during the replication of the genetic material. This repair system includes enzymes that proofread the DNA molecule after each replication and correct mistakes. Mistakes during the replication of DNA are caused by both molecules that occur naturally within the cell and by other agents, such as pesticides and radiation. The Human Immunodeficiency Virus (HIV) causes acquired immune deficiency syndrome (AIDS). The virus invades certain white blood cells, where it reproduces and destroys the cell. HIV is difficult to fight because every time it reproduces, it mutates slightly, and the immune system can not keep up with the changes. Today, scientists are using computers to design specific drugs, such as protease inhibitors that retard the growth of HIV, that will be more effective in controlling a microbe or neutralizing a toxin. The current use of numerous antibiotics to fight disease has resulted in the evolution of strains of bacteria that are resistant to antibiotics.

Learning Objectives

After studying this chapter, you should be able to:
(Other objectives may also be assigned by your instructor)

1.   Discuss the technology involved in genetic engineering.

2. Define a restriction enzyme and describe how it operates.
3. Describe some of the commercially available products resulting from genetic engineering.
4. Describe the technique of DNA fingerprinting, and discuss its importance in legal proceedings.
5. Discuss the problem of genetic diseases, and describe how gene therapy may be used to cure these diseases.
6. Discuss the relationship of defective genes to the formation of tumors and cancer.
7. Describe how DNA repair occurs between cell divisions.
8. Discuss how HIV virus manipulates the DNA in an infected cell to produce new viruses.
9. Discuss how computers are being used to design new drugs.
10. Describe how certain microbes have become immune to commonly used antibiotics.

Key Concepts

- Genetic engineering is a procedure by which foreign genes are inserted into an organism, or existing genes are altered to modify the functions of that organism.
  - Restriction enzymes are used to cut DNA molecules, exposing certain regions of unattached bases.
  - If the cut regions of two DNA molecules are complementary, the two molecules can be put together and the bases will bind and the strands will stick together.
  - If the new stretch of DNA spliced into the old DNA is a gene, it can now be expressed in the original cell.
  - This technique has been used to make insulin in genetically altered bacterial cells.
  - Genetic engineering has also been used to produce tomatoes with a increased shelf life and strawberries that are resistant to frost.
- The analysis of DNA in human tissue, a technique called DNA fingerprinting, is becoming increasingly important in the judicial system of the United States.
  - This analysis compares "variable number tandem repeats" in the DNA of two samples.
  - Segments of DNA containing VNTRs are spread on a gelatin-like material.
  - Radioactive DNA is then spliced to the pieces of DNA in the sample and a piece of photographic film is laid over the gel.
  - The procedure produces a photographic pattern of the analyzed DNA similar to a bar code.
  - The photographic patterns of different samples can then be compared to see if they both came from the same person.
- Cloning is the process of engineering a new individual entirely from the genetic material in a cell from another individual.
  - In 1997, Dolly, a sheep was the first mammal cloned in history.
  - Dolly was cloned by fusing an unfertilized egg, from which the nucleus had been removed, with a cell from an adult sheep.
  - The egg began to divide, and eventually produced an embryo and a lamb.
- Stem cell research involves totipotent cells taken from a very young embryo.
- Cancers are produced when genetic defects cause a cell's internal clock to fail, and the cell begins to divide without restraint.
  - Scientists have identified five or six genetic abnormalities that are responsible for cells beginning to divide without stopping.
  - Many of these DNA changes are acquired or accelerated by exposure to chemicals such as benzene, arsenic, atmospheric pollutants, and chemicals found in cigarette smoke.
- Genetic diseases may be cured by gene therapy.

- Some genetic diseases occur because a specific protein is not being made in an individual's cells.
- Gene therapy is used to replace a defective gene with a healthy one.
- Genes are injected into cells outside of the body, and the cells can then be introduced into the body.
- Current research efforts are aimed at identifying and producing "therapeutic viruses" that can be used to insert genes directly into the DNA of an existing cell in a person.
- Cancer is a disease that appears to result from the malfunction of the p53 gene, a gene that determines when a cell will divide. Cells that received healthy p53 genes through gene therapy ceased their uncontrolled divisions.
- Scientists have found it necessary to use double-blind clinical trials to test the results of new medical procedures. Using this technique, neither patients nor the physicians administering a drug know which group of patients receive the medicine being tested, and which group receives a placebo.

- Whenever DNA replicates, a certain amount of damage occurs to the DNA, either from the environment, or from metabolic byproducts in the cell.
  - A point defect is the substitution of one base for another in the double helix.
  - Mismatch errors may also occur when two bases on one side of the DNA stick to one base on the other side.
  - Enzymes patrol the length of DNA during replication to correct mistakes.
- Human Immunodeficiency Virus (HIV) causes the disease acquired immune deficience syndrome (AIDS).
  - The HIV virus invades certain white blood cells (T cells) in the human immune system.
  - The genetic material of HIV is composed of RNA.
  - Enzymes in the protein coat of HIV cause reverse transcription to occur within the cell, synthesizing a new DNA molecule from the viral RNA.
  - The new DNA is now inserted into the cell's DNA, where it directs the production of new viral RNA and enzymes to build new viruses that are released from the cell.
  - HIV is hard to combat immunologically because it mutates slightly every time reverse transcription occurs, producing slightly different strains of the virus.
- Scientists are now designing drugs with the aid of computers to specifically fit the three-dimensional shape of viruses and toxins.
  - Protease inhibitors that suppress the replication of HIV have been designed using this technique.

Key Concept: Gene Splicing

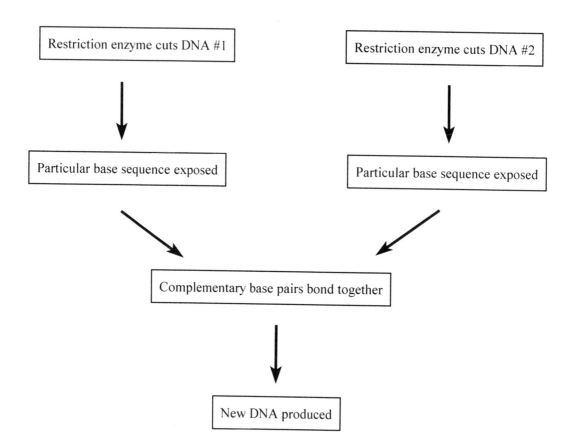

Key Concept: DNA Fingerprinting

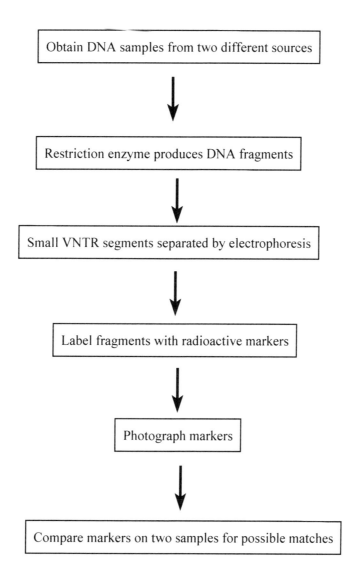

Questions for Review

Multiple-Choice Questions

1. Certain proteins that have the ability to cut a DNA molecule so that the DNA has several unattached gases at the cut end is a
    a. nucleotide
    b. molecule of RNA
    c. restriction enzyme
    d. virus
    e. none of the above

2. A human protein that is produced by genetic engineering of a bacterium is
    a. insulin
    b. saliva
    c. stomach acid
    d. intestinal juice
    e. tears

3. The analysis of DNA in human tissue is called
    a. genetic engineering
    b. gene splicing
    c. viral therapy
    d. DNA fingerprinting
    e. genetic therapy

4. A "variable number tandem repeat" in a DNA molecule is
    a. a region of repetition of a nonsense codons
    b. a region of instructions to make an amino acid
    c. a region of instructions to make a protein
    d. a region of instructions to make a lipid
    e. a region of instructions that corresponds to a gene

5. The standard technique in DNA fingerprinting relies on cutting DNA at how many different VNTR sites?
    a. 2
    b. 5
    c. 10
    d. 15
    e. 20

6. The process of replacing a defective gene with a healthy one is called
    a. genetic engineering
    b. DNA fingerprinting
    c. gene therapy
    d. viral engineering

    e.   viral therapy

7.   The term "in vitro" implies that a certain procedure
    a.   is being performed within a living cell
    b.   is being performed outside of a living cell
    c.   has yet to be performed
    d.   can never be performed
    e.   always involves gene therapy

8.   The term "in vivo" implies that a certain procedure
    a.   is being performed within a living cell
    b.   is being performed outside of a living cell
    c.   has yet to be performed
    d.   can never be performed
    e.   always involves gene therapy

9.   The p53 gene is involved in which disease?
    a.   diabetes
    b.   hepatitis
    c.   cystic fibrosis
    d.   AIDS
    e.   cancer

10.   In a double-blind clinical trial, the control group of individuals are given
    a.   a placebo
    b.   the experimental drug
    c.   both of the above
    d.   neither a. nor b.

Fill-In Questions

11.   Genetic engineering is the procedure by which _____ are inserted into an organism.
12.   _____ are proteins that have the ability to cut a DNA molecule so that the DNA has several unattached bases at the cut end.
13.   The DNA that is modified in genetic engineering procedures is located in the _____ of the cell.
14.   Using genetic engineering techniques, the colon bacterium, *E. coli,* is used to manufacture large quantities of the hormone _____.
15.   Through genetic engineering techniques, strawberries have been developed that are resistant to _____.
16.   The analysis of DNA in human tissue is called _____.
17.   A medical problem that arises because of the malfunctioning of genes is called a _____.
18.   At the present time, approximately _____ programs for gene therapy have been approved in the United States.
19.   _____ is a disease that occurs when a group of cells in the body reproduce without restraint.
20.   The disease know as AIDS is caused by the _____ virus.

There is no Crossword Quiz for Chapter 24.

Answers to Review Questions

Multiple-Choice Questions

1. c; 2. a; 3, d; 4. a; 5. b; 6. c; 7. b; 8. a; 9. e; 10. a

Fill-In Questions

11. foreign genes; 12. restriction enzymes; 13. nucleus; 14. insulin; 15. frost; 16. DNA fingerprinting; 17. genetic disease; 18. 200; 19. cancer; 20. human immunodeficiency (HIV)

Chapter 25

Evolution

Chapter Review

The diversity of life seen on the Earth today is due to the process of evolution, mostly through natural selection. The abundance and chronological distribution of fossilized remains of simpler forms of life in ancient rocks, followed by the appearance of more complex fossils closer to the present time is very strong evidence for evolution. Additional evidence comes from the fact that all life forms share certain basic biochemical mechanisms, especially with regard to the operations of DNA. Closely related species, such as humans and chimpanzees, also share most of their proteins. The presence of vestigial organs, such as the appendix in humans and tiny internal legs in whales, provide additional evidence of relationships between living organisms and relatives that are long extinct. Experiments in the laboratory show that complex organic molecules can be synthesized spontaneously from inorganic precursors. Scientists believe that these early organic molecules gained the ability to reproduce, and store information, in a manner similar to today's nucleic acids DNA and RNA. Somehow, these molecules became enclosed in a membranous envelope, and the first cells were formed. Current evidence indicates that life first appeared on Earth about 4 billion years ago. All subsequent life evolved from these early cells through a process called natural selection. Natural selection is based on two unescapable facts: every species produces more offspring than can possibly survive, and there is genetic variation among those offspring. Those individuals better adapted to the environment in which they live will survive longer, and, more importantly, reproduce better, thereby contributing more of their genes to future generations than do less well adapted individuals. Over long periods of time, organisms either change to adapt to changing environments, or they become extinct. The tempo of evolution and extinction is not a continuous process. Some eras are characterized by mass extinctions, followed by explosive evolution, while other eons are marked by little change in stable environments. Fossil and biochemical evidence indicates that human evolution extends back about 4.5 million years to a pre-human form called *Australopithecus*, a hominid that walked erect, but was not much larger than a chimpanzee. About 2 million years ago, the first tool-making hominid appeared. Other hominids evolved that learned to use fire, domesticate animals, and eventually develop agricultural practices. Modern humans, *Homo sapiens*, appear to have arisen about 200,000 years ago in Africa, and they migrated to all major parts of the Earth. Other human-like species are known from fossils, but their relationships to modern humans are still the subject of much scientific study.

Learning Objectives

After studying this chapter, you should be able to:
(Other objectives may also be assigned by your instructor)

1. Discuss the importance of fossils in understanding the evolutionary history of life on Earth.

2. Describe how similarities in biochemical pathways indicate relationships among organisms.
3. Cite the reason for the presence of apparently useless, vestigial organs in animals.
4. Describe the landmark experiments of Stanley Miller and Harold Urey in demonstrating that complex organic molecules could spontaneously arise from simple ones.
5. Discuss the possible route that led to the original formation of a cell.
6. Describe the process of natural selection and indicate how it can lead to the formation of new species.
7. Outline the major changes in living things over the last 4 billion years.
8. Discuss some of the causes of extinctions.
9. Describe some of the current hypotheses regarding the evolution of human beings.

Key Concepts

- Fossils are mineralized remains of dead organisms and their artifacts.
  - A dead organism may be buried, and, over a long period of time, the molecules in its body may be replaced by minerals, forming a fossil.
  - The fossil record refers to all of the fossils that have been found, cataloged, and analyzed by scientists.
  - Even though very incomplete, the fossil record can show changes in organisms over long periods of time.
  - The older the fossils, the more they differ from modern forms.
  - The Proterozoic Period (from 4 billion to 570 million years ago) is marked by prokaryote and simple multicellular fossils.
  - The Cambrian Period, approximately 570 million years ago, marks the appearance of numerous complex fossils, many with hard calcified skeletons.
  - The Paleozoic Period extended from 570 to 245 million years ago, and includes the development of all plant and animal forms.
  - The Mesozoic Period extended from 245 to 65 million years ago, and includes the age of dinosaurs.
  - The Cenozoic Period, the age of mammals, began 65 million years ago, and continues today.
- Scientists believe that the spontaneous formation of complex chemicals from simpler ones occurred prior to the appearance of living things.
  - Laboratory experiments proved that complex molecules, including amino acids and lipids, would spontaneously be synthesized in a mixture of hydrogen gas, water vapor, methane, and ammonia gas, if the mixture was supplied with an energy source simulating lightning discharges.
  - The concentration of organic materials in local places in the ocean produced a primordial soup.
  - Scientists hypothesize that the molecules in the primordial soup formed the first cells, but the exact mechanism is not known. The most promising idea results from our observation that polar characteristics of lipids causes them to spontaneously form small globules. Such globules would have enclosed small amounts of the soup, where further reactions could occur.
  - Because RNA acts both as a nucleic acid and as an enzyme, scientists hypothesize that it may have been the precursor of molecules like DNA that store great quantities of information very efficiently.
  - The first formation of a cell may have occurred in tide pools, or on rocks that were constantly splashed with the soup, creating highly concentrated mixtures that were exposed to the ultraviolet energy in sunlight.

- An alternative hypothesis is that the energy for formation of the first proteins and lipids may have come from heat produced at vents in deep ocean trenches.
- The first cells produced had no predators and no competitors, and probably multiplied rapidly.
- A meteorite from Mars shows evidence that life may also have evolved on that planet.
- Scientists believe that all life on earth today had its origin in early cells, and has resulted from natural changes brought about by natural selection.
  - Artificial selection is practiced by plant and animal breeders who select certain organisms for desired characteristics, such as speed in horses, taste in fruits and vegetables, and milk production in cows.
  - Natural selection occurs because of two basic facts:
    - Organisms produce more offspring than can possible survive.
    - There is genetic diversity among the individuals in any population.
  - Natural selection is the differential survival and reproduction of individuals in a population who are better adapted than others.
  - Natural selection leads to adaptation: structures, processes, or behaviors that help an organism survive and pass on its genes to the next generation.
  - Numerous examples of natural selection producing better adapter organisms exist, including peppered moths in England, and antibiotic-resistant strains of bacteria.
  - Even though many non-scientists do not accept evolution through natural selection because it does not agree with a literal reading of the Book of Genesis in the Bible, all of modern geology, biology and medicine are built upon evolutionary principles.
- Evolution through natural selection has resulted in a tremendous diversity of life on Earth.
  - Early life consisted of prokaryotes that spread through the oceans, perhaps very rapidly.
  - About a billion years ago, symbiotic relationships were set up that eventually led to the development of eukaryotes.
  - Individual cells began to come together to form colonies.
  - By about 600 million years ago, the seas were full of large multi celled animals and plants, many resembling jellyfish.
  - About 570 million years ago, the calcified skeleton appeared and there was an explosion in animal diversity.
  - Following the development of shells, animals invaded the land, and the insects are evidence of their success.
  - The internal skeleton characteristic of vertebrates appeared about 550 million years ago, leading to the development of the vertebrates.
  - Diversity of vertebrates includes fish, amphibians, reptiles, birds and mammals.
- Evolution does not always proceed at the same rate.
  - During some periods of time, organisms remain remarkably stable in a stable environment.
  - During these times, natural selection causes only gradual changes in organisms.
  - Mass extinctions occasionally occur, caused by natural cataclysms, such as volcanic eruptions, or collisions with asteroids from outer space.
  - Following extinctions, natural selection may cause extremely diverse changes in the survivors in geologically short periods of time.
- All scientific evidence indicates that humans arose due to the same principles of natural selection that operate on other organisms.
  - A rich fossil record traces human ancestry back approximately 4.5 million years to a small erect hominid called *Australopithecus*.
  - The genus *Homo* first appeared about 2 million years ago in Africa. This new hominid made

and used stone tools, and learned to use fire to cook food.
- Fossils of anatomically modern humans appear in rocks about 200,000 years old.
- Current evidence is that modern humans arose in Africa and spread throughout the world.
- Additional related populations have been identified in Europe as Neanderthal man, but their relationship to modern humans is unclear.

Key Individuals in Science

- In 1859, Charles Darwin (1809-1882) published *The Origin of Species*, a book in which he proposed the evolution of living organisms through the processes of natural selection.
- Alfred Russell Wallace (1823-1913) independently developed ideas about natural selection, and he and Darwin published a joint article prior to Darwin's book.
- Stanley Miller (b. 1930) and Harold Urey (1893-1981) performed laboratory experiments that showed that complex organic molecules could be synthesized spontaneously in a mixture of simpler gases if energy was supplied.
- In 1996, William Schopf discovered evidence for colonies of primitive bacteria in 3.9 billion-year-old rocks from Greenland.
- In 1980, Luis and Walter Alvarez, a father-and-son team, first proposed that the impact of a large asteroid killed off the dinosaurs and many other life forms.

Questions for Review

Multiple-Choice Questions

1. The scientific name for modern humans is
   a. *Homo habilis*
   b. *Homo erectus*
   c. *Homo neanderthalensis*
   d. *Australopithecus afarensis*
   e. *Homo sapiens*

2. A famous fossil called "Lucy" was a member of which species?
   a. *Homo habilis*
   b. *Homo erectus*
   c. *Homo neanderthalensis*
   d. *Australopithecus afarensis*
   e. *Homo sapiens*

3. Which of the following represents a vestigial organ?
   a. thumb
   b. appendix
   c. eye
   d. stomach
   e. elbow

4. One of the universally possessed proteins that has been analyzed extensively for biochemical evidence for evolution is
   a. DNA
   b. RNA
   c. cytochrome C
   d. enzymal RNA
   e. amino acids

5. Punctuated equilibrium is a theory of evolution that states that
   a. not all organisms follow the same rules of natural selection
   b. all animals evolve at the same rate
   c. all species live only a certain number of years, punctuated by extinction
   d. evolutionary changes usually occur in short bursts, separated by long periods of stability
   e. some species of organisms never change

6. Dinosaurs disappeared approximately
   a. 50 million years ago
   b. 65 million years ago
   c. 100 million years ago
   d. 200 million years ago
   e. 250 million years ago

7. Which geologic period is known as the age of mammals?
   a. Proterozoic
   b. Paleozoic
   c. Mesozoic
   d. Cenozoic
   e. none of the above

8. Which geologic period is known as the age of reptiles?
   a. Proterozoic
   b. Paleozoic
   c. Mesozoic
   d. Cenozoic
   e. none of the above

9. Life first appeared on Earth approximately
   a. 10 billion years ago
   b. 5 billion years ago
   c. 4 billions years ago
   d. 3 billion years ago
   e. 1 billion years ago

10. The author of *The Origin of Species* was
    a. Alfred Russell Wallace
    b. Charles Darwin
    c. Harold Urey
    d. Luis Alvarez
    e. Stanley Miller

Fill-In Questions

11. Organisms that lived in the past, but have disappeared forever from the Earth are said to be
    _____.

12. The Urey-Miller experiments combined a mixture of water vapor, hydrogen, methane, and
    _____ in a flask.

13. Scientists hypothesize that organic molecules accumulated in the early ocean for form a primordial
    _____.

14. The first kind of cell to evolve was a _____ cell.

15. The mechanism for evolution of living things as proposed by Darwin is _____.

16. Any structure that helps an animal survive and pass on its genes is called an _____.

17. Cyanobacteria are single-celled life forms that produce oxygen as a byproduct of _____.

18. The Cenozoic period is known as the age of _____.

19. The longest geologic era is the _____.

20. A stone replica of a once-living organism is called a _____.

Crossword Quiz: Evolution

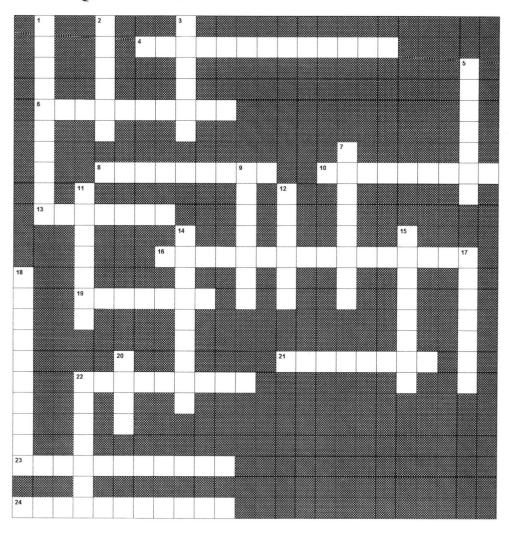

ACROSS

4. Single-celled life forms that produce oxygen as a byproduct of photosynthesis
6. This type of selection is practiced by plant breeders
8. An organ with no useful function
10. The geologic time period that precedes the Mesozoic
13. The first species of human to use fire
16. The oldest genus hominids that walked erect
19. This type of selection leads to evolution
21. The age of reptiles
22. Change in living organisms over time
23. These humans thrived in Europe 50,000 years ago
24. The earliest geologic period

DOWN

1. A cell that does not contain a nucleus
2. A mineralized replica of a once-living organism
3. He published The Origin Of Species
5. This species of humans was the first tool-maker
7. This geologic period is noted for the appearance of the hard external skeleton
9. He proposed that the collision of an asteroid caused the extinction of the dinosaurs
11. An organic gas that was probably present in the atmosphere of the early Earth
12. He conducted experiments on chemical evolution on the early earth
14. This type of cell possesses a nucleus
15. The age of mammals
17. The species name of modern humans
18. Any structure that increases an organism's chance for survival
20. The genus of human beings
22. An organism that has forever disappeared from the Earth

Answers to Review Questions

True-False Questions

1. e; 2. d; 3. b; 4. c; 5. d; 6. b; 7. d; 8. c; 9. c; 10. b

Fill-In Questions

11. extinct; 12. ammonia; 13. soup; 14. eukaryote; 15. natural selection; 16. adaptation; 17. photosynthesis; 18. mammals; 19. Proterozoic; 20. fossil

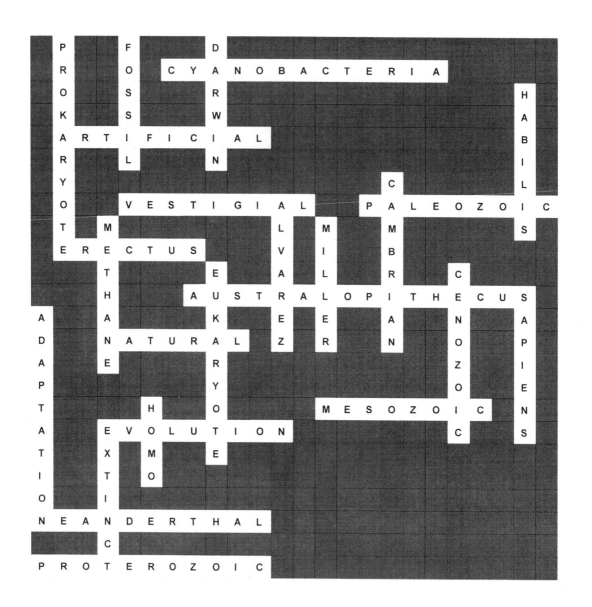